05/0

Buscando información
mejorar mis oportunidades de conseg
de "Manager Manufacturing Engineering" me en
éste libro.

Kairo Z. Hannon

OLD-FASHIONED ENGINEERING & MANAGEMENT

*The Only Acronym
You'll Ever Need*

OLD-FASHIONED ENGINEERING & MANAGEMENT

# *The Only Acronym You'll Ever Need*

## A Simple and Practical Guide to Ensuring Success In Business

Mark A. Ousnamer, M.S., P.E.

ENGINEERING
& MANAGEMENT
PRESS

*Printed in the United States of America.*

00    99    98    97    1    2    3    4    5

LIBRARY OF CONGRESS
**Cataloguing-In-Publication Data**

Ousnamer, Mark A., 1961–
    OFE&M: the only acronym you'll ever need:
a simple and practical guide to ensuring success in
business / Mark A. Ousnamer.
        p.    cm.
    Includes bibliographical references.
    ISBN 0-89806-175-X
    1. Industrial management.  2. Engineering
design.  3. Success in business.  I. Title.
    HD31.0844    1997
    658--dc21                                        97-11359
                                                          CIP

ISBN 0-89806-175-X
ENGINEERING & MANAGEMENT PRESS
*Director:* Eric E. Torrey
*Editor:* Forsyth Alexander
*Designer:* Candace J. Magee
*Publishing Assistant:* Matthew Orsagh

**EMP** *25 Technology Park, Norcross, GA  30092*

# Contents

# Preface

I HAVE SPENT A LOT OF TIME WONDERING how to phrase the jumble of thoughts and ideas that have been banging around in my head for the last few years. Then, I was invited to speak on Old-Fashioned Engineering and Management (OFE&M), and I had to make a decision about my beliefs and whether I could communicate my philosophy effectively. I was forced to devise a concise definition because too many people were losing interest in my soapbox harangues. They felt I probably had a point, but they couldn't find it. The words below are the result.

Old-Fashioned Engineering and Management is the philosophy that anyone is capable of being more effective in their endeavors simply by focusing on and utilizing the skills and knowledge they already possess.

*Corollary #1:* Be creative in applying your skills and knowledge.
*Corollary #2:* To be most effective, identify areas for improvement and seek to strengthen them whether they are personal, educational, or experiential.

That's it. That's all there is. The contents of this book summarized in a few carefully chosen words.

Over the years, I have used many methods to teach difficult concepts to people with diverse backgrounds. I know that people learn at various speeds and relate to information differently. I usually try to simplify difficult concepts so that I can understand them piece-by-piece. Therefore, I am going to present the basic principles of OFE&M several ways.

I am not a writer by trade, but I enjoy writing. I hope I am being creative in this presentation. I chose to use a first-person narrative because I want to be different and personal. In college writing classes, we were told never to write in the first person unless telling a story. In a manner of speaking, OFE&M is a story. It is the story of how my experiences created a belief system I feel is important enough to record and communicate to others.

# Introduction

IT IS BECOMING INCREASINGLY DIFFICULT to survive in today's business environment. Few, if any, engineers or managers at any corporate structure level will argue this point. It is currently a "given" that companies or businesses must now fight for a share in a world market that does not acknowledge geographical boundaries or national loyalties. Companies now need answers to questions that just recently could safely be given passing notice and then "back-burnered" without regard to the consequences. After all, if problems arose, priorities were shifted, and the most pressing concerns were dealt with as needed.

In today's business environment, that problem-solving approach will, more often than not, put most of a company's employees squarely in the unemployment line. The competition is becoming more and more efficient at dealing with the problems of doing business. In fact, they are turning these problems into golden opportunities and enlightening challenges.

In response to this new "challenges and opportunities" quagmire, many companies are turning to the experts for advice. New methodologies, ideas, and techniques are being taught, trained, implemented, invoked, designed, applied, and administered in hi-tech and imaginative ways. At the root of this movement is the concept that to deal with the changing business world, complex technological tools and methodologies are necessary.

Alphabet soup titles, seminars, and top-dollar consultants are the cornerstones upon which successful businesses are apparently built today. Regardless of the industry, the myth is perpetuated that a guru with the proper technique is what it takes to survive, as long as the technique is appropriately complex and incomprehensible. In-house personnel are allowed brief moments of insight, but only if it raises more questions than answers and does not lead to an epiphany.

Therefore, I am staking out the following mountain top. I am putting the turban of wisdom on my head and wrapping myself in the mantle of Old Fashioned Engineering and Management. Thus cloaked, I propose that the path to corporate survival and global good health need not be fraught with complex technological dangers. It can actually be rather simple and straightforward in practice. OFE&M is based upon the fundamental clichés "there is nothing new under the sun" and "keep it simple, stupid." These clichés are then coupled with common sense (generally known to be rare in practice), a little human insight, and empathy,

My viewpoint is a multi-faceted one. I have worked as a general laborer, a salesman, a mechanic, a production line worker, a production line supervisor, a draftsman, a junior engineer, a middle manager, a business owner, and a consultant. I have taught engineering labs at the university level and seminars on various specialized topics. This experience and the people who have taught me what I know (directly and indirectly) were the impetus for my position on the mountain top.

This book is not a treatise on how you shouldn't attend seminars or listen to experienced experts. It is a mechanism I hope will encourage engineers and managers to understand better and utilize the basic abilities they have. Often, we spend inordinate amounts of time and money chasing knowledge we believe will help us address the problems we face, when, in fact, we already have all the tools we need.

I am writing for all those engineers and managers who, over the years, did not argue with me about OFE&M. Instead, they agreed with me and contributed yet more facts, figures, and stories to support my theory. I have included some of these in homage to those frustrated (and sometimes thoroughly amazed) individuals who have worked for and with "The Others." "The Others" are engineers and managers that live and breathe theory, believe gut feelings are simply indigestion that will pass, and readily admit to living in an ivory tower.

These "lies and war stories" are integral to OFE&M. They relate concepts and ideas better than dry facts and figures. I confess I have also used them so that I didn't have to research many facts and figures. I plead early senility if it seems I have stretched the numbers from personal anecdotes or if someone figures out that "Company A" really does exist and the "The Other" in question was an old boss or coworker. In most cases, I do not think I could create better examples than the ones life generously hands out. I am also sure that each of you has at least one story worthy of inclusion in this book.

# Acknowledgments

MANY THANKS TO MY MENTORS over the years. First and foremost to Jack Ousnamer, my father, who first instilled the roots of OFE&M before I had a clue what it meant. I also offer thanks to Dr. John Imhoff who showed me that engineering, academia, and "being a people person" are not mutually exclusive. Then there are Phil Warner, who taught me the ropes of working for large corporations despite my inability to integrate his teachings fully and Don Dye, who took over guardianship when I started my own business.

Thanks also to my friends and compatriots Soma Coulibaly, Stacy Hope, Joe Rogers, Kim Wilcox, and Russ Hawkins for their input on the manuscript. And to Eric Torrey and Forsyth Alexander of Engineering and Management Press for their hard work and positive attitudes in making a concept reality.

Finally, utmost appreciation and thanks go to my wife Pamela for her support of yet another adventure I have undertaken.

# Dedication

*In memory of Mildred Porch, my grandmother, who was always unwavering in her support and gentle with her criticism.*

OLD-FASHIONED ENGINEERING & MANAGEMENT

*The Only Acronym*
*You'll Ever Need*

# CHAPTER 1

# What is OFE&M?

> OFE&M is the philosophy that anyone
> is capable of being more effective in their
> endeavors simply by focusing on
> and utilizing the skills and knowledge
> they already possess.

*Corollary #1:*  Be creative in applying your skills and knowledge.

*Corollary #2:*  To be most effective, identify areas for improvement
and seek to strengthen them, whether they are personal,
educational, or experiential.

IF LEONARDO DA VINCI WERE ALIVE TODAY, he might be confused by
the way we practice engineering in this day and age. Leonardo's greatest asset was his mind. He was constantly thinking and often just day-
dreaming about what could be accomplished by trying new approaches to problems. Interestingly, he thought of himself first and foremost
as an engineer—despite his other accomplishments. Generally, his
ideas and concepts were not new. Instead, he refined the designs of
others and the technology used to fabricate those designs.

I enjoy examining his sketches and imagining what it must have been like to propose the outrageous ideas (people flying!) that he regularly conceptualized and empirically tried to turn into reality. Leonardo did not have a huge research and development budget—quite the contrary, even if only a portion of the history books can be believed. A budget is a budget, whether it is presented to a patron or to a board of directors. He certainly did not have the latest engineering technology in a laptop he could reference at the press of a button nor did he have access via phone, fax, or modem to the leading experts of the day. Despite all these "handicaps," he is one of history's greatest engineers. He defined basic facts and theories and then built on them with creativity and imagination. Leonardo was an old fashioned engineer. OFE&M requires creativity and a willingness to embrace the unknown.

The best engineers and managers I have known personally rely first and foremost on their own instincts and experience to ensure their own success, the success of their departments, and ultimately that of their companies. They accept that the basic tools of their profession must be mastered before they can apply them, yet have not let formal training overshadow their instincts. OFE&M requires knowledge in a field, but it also relies heavily on developing and trusting instincts.

Engineers and managers of any period, nationality, or company size share the same basic professional tools. For engineers, the tool kit might include an understanding of physics, materials, chemistry, statics, drafting, and any other knowledge common to all engineering disciplines. Managers generally consider cost accounting, sociology, psychology, and marketing as bare essentials for survival. How these basic tools and others are applied and utilized to achieve desired results is the heart of OFE&M. No distinction is made between degreed engineers and managers and those who made their way into their positions through hard work and experience. Both schools, the one of hard knocks and the one of formal universities, have good points. They both have their weaknesses, too. OFE&M requires an understanding of both strengths and weaknesses.

I once worked in a production facility making deep drawn stainless products. This facility was encountering low yield and poor quality in a small-batch production run. The scrap rate was high enough that I was asked, as a degreed engineer, to help determine what caused the defects. The initial basic investigation determined that the draw being performed in the first operation was beyond theoretical design limits. The production manager's response to my insightful discovery was, "I've been told many times by the steel mill engineers that the draws we perform are impossible and that they will not take any responsibility for our scrap rate. They say we are not supposed to be able to get a single good part from this operation. Now have you got any ideas why it's not working this time? I've got production to get out."

Experience had taught the production manager that the operation could be successfully performed, but not how to troubleshoot the operation with timeliness and efficiency. However, his intimate knowledge of the process, when combined with my experimental design and statistical analysis techniques, allowed us to determine the problem's probable cause and address it.

Perhaps the simplest way to define the full scope of OFE&M is to imagine a small business run by a single owner. Let's assume the owner has a successful manufacturing company selling "Neat Stuff." Small in this instance is a company of 15 employees with gross sales less than $1,000,000. The owner is responsible for all major decisions that affect financing, product design, production, staffing, and customer satisfaction. Two hats, those of "engineer" and "manager," fit on the owner's head interchangeably and can be switched at a moment's notice.

This business owner's decisions have an immediate impact on the lives of the company's employees and the company's bottom line. There is no fifteen-layer management buffer, nor can mistakes be hidden within the workings of a multi-divisional, multi-million dollar corporation. If the owner takes a misstep in any operations area, there is the potential for devastating results. This type of business model allows you to ascertain the probable results of action and reaction without the damping effect and delays inherent in a more complex system. Imagining how The Neat Stuff Co. would address issues will, I hope, provide you with a clear, simple idea of how the tenets of OFE&M can be applied successfully in any organization. (Now is not the time to try convincing me that wonderful and intricate simulation software will answer all the questions. I will address simulations at a later time.) OFE&M requires making decisions and taking actions as if the results will have an immediate impact upon personal health or wealth.

> OFE&M requires making decisions and taking actions as if the results will have an immediate impact upon personal health or wealth.

When I say an immediate impact upon health or wealth, I am not referring to abstract concepts or textbook theories. I mean literally. I experienced firsthand two incidents that have indelibly marked that concept firmly in my memory. The first involved a woman that worked in a department I supervised. We had an old mechanical press for crimping metal parts that was slow but effective. The day after Christmas I was in the plant (while everyone else was on vacation) running prototype parts on the press. I wanted to dial-in the tooling before the "after-the-holiday" production started. After a few hours, I went home. The next time the press was operated, a dog pin that controlled the press cycle broke and several of the operator's fingers were amputated. My office was next to the department, and she came to me for first aid. My first aid training notwithstanding, it was not an easy injury to deal with.

My investigation into what caused the part failure uncovered the company vice president's decision to have a $250 special-order machine part made locally for $50. The local vendor's materials and quality assurance procedures were inferior, and the pins made over the years had shown a tendency to fail quickly and frequently. When the pins failed (broke in two parts), they had to be replaced. Production parts were scrapped, and production stopped while replacement pins were installed. The vice president of manufacturing considered this downtime a cost-effective alternative to spending an extra $200 (per pin) to prevent catastrophic failures. His decision to use the bootleg parts ultimately cost an operator three fingers on her dominant left hand. By my calculations, I missed losing my fingers by fourteen press cycles the day I was in working on prototypes. I am also left-handed. The cost of replacing the mechanical press with a hydraulic press was deemed negligible once I discovered the reason behind the failure.

A year and ten days later at the same company, the same vice president put a grinder operator to work on one of the big hydraulic forming presses. A manpower shortage for the shift prompted the VP to make the assignment to keep up production. The grinder operator had never been trained on the operation and safety procedures of a hydraulic press. While the operator was performing a tooling change, the press drifted and reset, amputating his thumbs. Once again, as the plant safety officer, I was called to administer first aid and investigate the accident. No matter how many times you have to deal with the pain and suffering caused by bad management decisions, it never gets easier.

An OFE&M manager is aware of the impact his or her decisions can have. If a decision requires expertise beyond his or her understanding, he or she seeks any assistance necessary to ensure a viable and responsible course of action.

OFE&M can further be defined by a few time-honored and well-worn clichés that will be covered in some detail in Chapter 6.

# CHAPTER 2

# Characteristics of the Old Fashioned Engineer and Manager

If you received a formal engineering education at an ABET accredited school as I did, you may have noticed there was no room in the curriculum for developing the human characteristics necessary in the social situations that are part of daily job performance.

I HAVE STUDIED MY SHARE of formal management and engineering texts over the years in college and after as suggested reading by different colleagues. (I even took a college course specifically geared toward management of engineers.) What is never explained in plain English are the characteristics or traits that make you become a successful engineer or manager. Invariably, the text deals with various techniques and tools. Even the organizational behavior courses and texts seem to focus on situations or case studies that may arise, while suggesting typical approaches for handling said situations. Nary a word is said about the personal traits that could help you with these situations.

Psychology and sociology both emphasize the range of traits and characteristics individuals exhibit as they perform their jobs or go about their lives interacting with others. Although the management texts may point them out as things to expect from others, where is the list of traits to cultivate within yourself as you go about performing your duties as a manager? If you received a formal engineering education at an ABET accredited school as I did, you may have noticed there was no room in the curriculum for developing the human characteristics necessary in the social situations that are part of daily job performance. No wonder engineers are stereotyped as mechanical, unfeeling, thinking machines, and managers (or bosses, to use a common term) are seen as self-centered, self-serving, company-bottom-line hardcases. Most of us can think of one or two exceptions, but they are few and far between.

Engineers and managers should be honest, creative, communicative, documentation-oriented, and be able to admit to their mistakes quickly and comfortably.

*Creativity is needed to prevent grooves that work well in a few situations from becoming ruts that don't work at all.*

Honesty is self-explanatory. If your word cannot be trusted and those people that work for and around you must constantly verify the validity of your statements, you will become mired in distrust. You will be unable to perform at optimum levels.

Communication goes hand in hand with honesty. Learn to write effectively. Clear communication—listening and speaking—eliminates confusing gray areas and speeds any process. There is an old saying that goes like this: "God gave us two ears and only one mouth, so which one do you think is more important?"

A good example was related by an instructor teaching a Hazwoper (Hazardous Waste Operator) class I attended. He was explaining that good communication can improve safety on any given work site because it helps identify areas that need improvement. At one time, he was in charge of safety at a work site that had a high head-injury rate. He was told that the workers "refused" to wear the mandatory hard hats when tying rebar for concrete forming. When he spotted a worker without his hard hat, he didn't yell and scream the latest Worker's Compensation figures at him. Instead, he calmly asked the operator why he wasn't wearing his hard hat. The operator responded that the hard hats the company purchased didn't have chin straps. For most jobs in the company, chin straps weren't necessary and workers refused to wear them. The company decided to buy hard hats without chin straps, thus saving money. Unfortunately, a hard hat cannot stay in place without a chin strap when a worker bends over to tie rebar. The workers were not refusing wear the hard hats; rather, the hats simply kept falling off. A timely purchase of chin straps solved the problem.

Communication sometimes means using language or imagery that is effective in its simplicity. While trying to emphasize to a line worker that in-process parts must be treated as valuable items that should not be damaged lightly, I knew that quoting reject and rework percentages and accompanying revenue losses wouldn't communicate my message as clearly as I wanted. On the other hand, a $20 bill has a very definite universal value. So, I told her that, by the time the part got to her work station, it was worth about $20. I then pulled a $20 bill from my pocket (a minor miracle at that stage in my career), and pretended to tear it up and throw it on the floor.

Product redesign, process evaluation, and training restructuring resulted in rework and rejection percentage reductions, from almost 40 percent to less than 1 percent. The effectiveness of my communication was driven home eight months later when the same operator almost dropped a part on the floor in front of me (which would have ruined it). Glancing up at me quickly while protectively cuddling the part, she laughed nervously and said "I know, I know. Don't say anything. It's just like tearing up a $20 bill if I damage it!"

The entire department shared a laugh over my facial expression and the way she was cradling a stainless steel part like a little baby. I had effectively communicated an important message about quality, and it was easily remembered without negative connotations months later.

Documentation is critical to any operation. As the old saying goes, it greases the wheels of the machine. Proper documentation saves rehashing old ground on revised projects and can be used as a quick reference when clarification is needed. Who hasn't had a project pushed back a couple of months or years, only to have it thrust to the forefront of the impending doom list? Documentation is the only way you can get back up to speed efficiently and effectively.

I have also used documentation as a tool to gain credibility when requesting capital appropriation dollars. I wanted to purchase a new vertical milling machine with digital readouts for my maintenance department, and I had estimated a fourteen-month cost payback. With some teeth pulling, it was approved and purchased. Rather than simply walking away from a successful project (after all, I had gotten the approval and bought the mill), I considered the project only half completed. I placed a log book next to the mill in the maintenance department and required all work on the mill to be logged by type and time spent running the mill. In nine months, I had concrete proof that the mill had paid for itself by doing machining in house that we had previously been sending out. It was easier to get my capital budget items approved after that. Also, the complaints about the log book faded when it became apparent how much easier it was to purchase new tools after our little documentation exercise.

Try to take the road less taken. True, it's the one over-grown with weeds, but it will allow shortcuts and quantum leaps in understanding even the most mundane daily tasks. If you push the envelope and try something new, expect to stumble occasionally and even fall flat on your face once or twice. Mistakes happen all the time; you are either inhuman if you don't make mistakes or you are deluding yourself, so learn to deal with them expediently. Mistakes cannot be corrected until they are identified.

The most embarrassing mistake I made in recent memory was when I took time to go to the gym and work out one morning when I believed my schedule allowed it. When I returned to the office around 10:00 a.m., I had an urgent message to call a client. My call was greeted with a laugh and "Did you have a good work out?" Even though my relationship with this particular company was amicable, I grew more suspicious about the call's tone as it progressed without mentioning a business-related topic. My suspicion turned to discomfort when I was asked if I remembered a meeting scheduled for 9:00 a.m. that day. Discomfort turned to red-faced embarrassment when it was pointed out that I had called the meeting. I admitted that I had completely forgotten the meeting and apologized profusely. I'm glad I owned up to my mistake, because the next statement complimented me on my very well-written memo to the company's management and staff. Included in this memo were the topic, date, and time for the meeting I had missed. Several weeks before, I had conducted training on written and oral communications, so I figured I might as well graciously and

> Mistakes happen all the time; you are either inhuman if you don't make mistakes or you are deluding yourself, so learn to deal with them expediently. Mistakes cannot be corrected until they are identified.

**11**

humbly accept the compliment and laugh along with the client. Rescheduling the meeting was relatively simple (I wrote it in bold letters in my daily planner this time), and no harm was done. I am still teased about the episode from time to time, but the fact that I am still doing work for that company five years later speaks for itself.

On the other hand, I am often in a position to witness honest mistakes that result in confusion because they are not acknowledged. In at least one case, "ducking" the issue could have cost the company involved many thousands of dollars.

I was asked to order some expensive specialty hand tools from Germany for a client because I speak German. I don't exercise my language skills that often, and I thought I misunderstood or mistranslated the vendor's statement that the hand tools had already been ordered. It seems a company engineer had ordered two of the tools several weeks earlier, without completing all the proper paperwork or notifying the purchasing agent.

By directly dealing with mistakes positively, you can usually identify, correct, and successfully modify a project in the time most people take to bury their mistakes.

After several phone calls, we finally received verification in English (my contact's English was better than my German) that, based on the supplier's phone records, the tools had indeed been ordered for the local plant. We decided to lessen the number of tools on the order I was handling at that time. Based on the exchange rates at the time, it amounted to a savings of more than $15,000. The engineer in question never did admit to going outside the normal procedure for ordering high-dollar items, and the time we spent ferreting out answers could have been better spent elsewhere.

By directly dealing with mistakes positively, you can usually identify, correct, and successfully modify a project in the time most people take to bury their mistakes. A mistake is a lot like a festering wound. Placing a clean bandage on an infected wound without treating it properly makes it look neat and clean for a short while—until the bandages soak through and draw attention or the patient dies. Either way, not admitting to your mistakes is a losing proposition.

# CHAPTER 3

# How To Implement OFE&M

As a practitioner of OFE&M,
every decision you make
should consider the elements of safety,
quality, and productivity at the individual,
department, and corporate levels.

OFE&M IS A STRAIGHTFORWARD PHILOSOPHY at its basic levels. Despite its conceptual simplicity, however, implementing it can be a personal struggle. Each individual undertaking OFE&M must take responsibility for his or her actions, look hard and true at motivating factors in his or her career and life, and honestly identify weaknesses that must be strengthened. If you fail at implementing OFE&M there is no one to blame but yourself. There's a flip side, though: if you are successful, you get the credit for taking the higher road and making it work.

I have found that I like conceptual groupings of three, so I have created a *Rule of Threes Matrix* (RTM) and a *Rule of Threes Statement* (RTS). The RTM groups the traits of education, experience, and personality against the levels of individual, department, and organization in the traditional rows and columns of a matrix. The *Rule of*

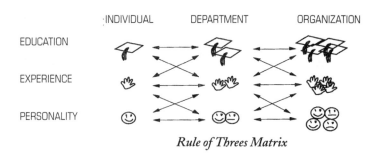

*Rule of Threes Matrix*

*Threes Statement* says that the interaction of three traits—education, experience, and personality—with three levels—individual, department, and organization—is three times as confusing and three times as important as anything else you will try to make sense of in your career.

To make OFE&M work at its lowest level—the personal level—you need consider little more than your own circumstances. This includes how much formal education you have, the type of personal and work experience you can call upon, and what type of personality you have. Each should be examined honestly to determine whether it needs to be enhanced, modified, expanded, or changed outright. It is possible that one or more may need absolutely no effort applied at the individual level. This is viable if you are being honest about yourself and what you perceive is necessary for success.

Education. For you to have enough formal knowledge to achieve your goals, is a degree necessary? If you are in management, would an MBA truly be useful, would it be politically correct, or does it simply look good? Would that specialty degree give you the information and confidence to perform more efficiently? Or, can workshops and specialty seminars provide the information you need and want? Should you consider an internship under someone with specialized knowledge?

Education comes in many shapes and forms. The questions you ask yourself before seeking educational avenues can make or break you in the long term. I have a friend whose wife has three separate degrees (I know accounting and nursing are two and I think a general business degree is the third), yet she is considering more education because none of the career paths open to her at this time appeal to her. She could be experiencing professional student syndrome, or she could be listening to counselors and not her own mind. The result is the same—a lot of education not being utilized to its full potential. Education cannot be wasted, as diverse knowledge can be combined in unique ways into creative solutions. Having a plan for seeking out knowledge can prevent you from spending lots of money for a delayed return on the investment.

> Ignorance is not a crime. In fact, I learned that open acknowledgment of my ignorance was a wonderful way to learn new information.

The easiest way to determine if and what kind of further education you need is to ask yourself how many times during the course of a project (or even during a day) you find yourself wishing you knew more about a particular topic. If you frequently are irritated or even caught flat-footed without answers to questions or situations you feel you should be able to handle, consider expanding your educational base. Ignorance is not a crime. In fact, I learned that open acknowledgment of my ignorance was a wonderful way to learn new information.

While still enrolled in the college of engineering, I applied for a drafting position at a local machine shop. My job was to reverse engineer parts and create file drawings the machinists could use in building replacement parts for the various clients' machinery. I did not have a machining background, and the machinists were quick to give the "school boy" definitive lessons that pointed out my educational shortcomings. The first was how to hold a micrometer correctly to take a reading without dropping it. The good news was I realized it wasn't a C-clamp; the bad news was I almost damaged the part and the micrometer (mic) because my juggling skills aren't all that great, and I almost dropped both. I was quickly taught to hold the mic frame in my palm and securing it with the pinkie and ring finger. The spindle then can be turned smoothly with the thumb and index finger of the same hand, thus freeing the other hand to control the part being measured. I appreciated the lesson I learned that day, despite the good-natured ribbing I had to endure. Thereafter, if I had a question about a process, all I had to do was step out onto the floor and say to the nearest machinist, "Excuse me, can I show my ignorance for a moment?" The reply was always, "Why, of course!" punctuated with a smile and a ready answer. I have always felt that any day is a good day if I learn something new.

> Experience is often the best education you can receive. It is generally low-cost and imprints lessons indelibly on your memory.

Experience is often the best education you can receive. It is generally low-cost and imprints lessons indelibly on your memory. My boss at the machine shop "offered" me valuable die design experience by salvaging machine bolts and tool steel from some 100 obsolete die sets while tearing them down over the course of a week. I learned how die sets are designed, machined, and assembled. I also learned that precision machining parts assembled with close tolerances are very sharp and should be handled with care. This is common sense if you work with die parts for any length of time, but I can't remember a textbook ever mentioning this practical consideration. Classroom teaching rarely achieves the same level of retention as hands-on experience. Classroom education is generally expensive both in dollars and time invested without monetary return, but you can often be paid to gain experience. The downside to experience (and many will argue that it is a positive, not a negative aspect) is that it also means a big time investment. It cannot be gained any other way.

Granted, some of us have gained turbo-instilled experience by having high-profile and short-deadline projects thrust upon us. As a rule, however, unless you seek out this type of experience, it generally goes to those individuals who thrive on pressure. I gained much of my experience this way: I volunteered for the projects everyone else in the department shied away from. It wasn't until I was older (and had gained some of that valuable experience) that I learned that "high-risk-high-reward" applies to gaining experience as well as monetary investments.

In many ways, experience is an investment of a little knowledge applied over time to yield more knowledge, just as monetary investments in long-term stocks can yield great financial rewards. Like playing the stock market, though, you must be careful. Many good managers and engineers have jumped onto a cutting-edge project with high risk, hoping to make a name for themselves and gain unique and marketable experience, only to find themselves on a dead branch too rotten for fire wood and about to be pruned by the corporate accounting department.

Movers and shakers make a lot of multi-million dollar mistakes, but we don't always hear about them because they are also adept at correcting their mistakes and pulling the fat from the fire.

The best advice I can give about gaining experience is from a conversation I overheard between two businessmen on a long and boring flight. The younger man, new to his company, was lamenting the fact he had not yet found the best way to gain "good" experience. The older gentleman made the following statement, which I found to be hilarious at the time. "The best experience that you can get is to find a mover and a shaker in your industry, someone who deals in high-dollar, high-risk deals, and watch what he does, especially when he makes a multi-million dollar mistake. It costs you nothing, yet you get full benefit from the experience, and it doesn't hurt your reputation." I was caught eavesdropping, because I literally burst out laughing at this little jewel. How cold-blooded can you get, taking advantage of someone else's mistakes for your own personal gain? Then, I realized the full truth of the statement. Movers and shakers make a lot of multi-million dollar mistakes, but we don't always hear about them because they are also adept at correcting their mistakes and pulling the fat from the fire, in a manner of speaking. What better experience can you get than that?

Whether you understand why I found the older businessman's advice funny or not depends on your personality. If you have never taken a Meyers-Briggs personality profile, I strongly recommend that you do. Actually, any standard personality test that is administered by an objective person will do. The idea is to get a feel for how you interact with others, handle decision making, and approach information gathering. If you are going to maximize your career potential and your effectiveness, you must understand the basics of how others perceive you. Humans are social animals. Personality traits that are understood and addressed positively will get you much further than blindly being frustrated project after project despite "doing all the right things."

*Effectiveness as a manager or engineer is as much about people skills, self-awareness, and the understanding of multiple interaction levels as it is about knowledge and how that knowledge is applied.*

This is not to say you must undergo serious behavior

modification if you discover that your personality is not "ideal" for your position. A quiet and introverted manager can be perceived as ineffective and unsupportive. The same individual can also be positively perceived as polite, reserved, and allowing his employees initiative. I personally have been described as assertive, aggressive, and a real go-getter. I have also been described as abrasive and obnoxious. The difference in perception is how aware you are of your personality traits and how your actions are received by others. This leads to the next levels of the *Rule of Threes Matrix:* department and organizational traits and interactions.

Departments and organizations have distinct education levels, experience levels, and personalities. Corporations are even granted legal status as entities, and it is easy to find reference materials related to corporate culture that correlate directly with an individual's personality.

Awareness of a department's characteristics will make or break you as an individual within it. Consciously trying to modify a department's characteristics as a manager is one thing, but swimming upstream as a member of the department is another. I have a great deal of experience with this topic—I have been fired or asked to leave every position I have ever held within a corporate structure. (This is why I feel qualified to write about the consequences of ignoring the *Rule of Threes Matrix.*)

Many friends from early in my career openly declare how much they like me as a person and how much they hated working with me as a young engineer and manager. While reminiscing about my early days as a manufacturing department's junior project engineer, one friend even recently admitted that my attitude about always being technically right was incredibly obnoxious. He started laughing at some mental image he had conjured up and blurted out that the fact I was probably right actually had made it worse. So much for my early social charm and business manners.

Political suicide in business comes in many forms and I exercised most of them. Casually throwing out answers to the difficult questions posed by corporate executives in high level board meetings when you are a junior engineer is not smart. I always managed to time my responses immediately after a middle manager (my own or another department's) promised to investigate and come up with a response in a few weeks. I once did this in a small meeting between my immediate boss, the plant operations manager, and a visiting senior corporate vice president. The VP had a reputation for firing people on the spot for giving him bad information or trying to make themselves look important.

I uttered my definite response to this VP's question. My boss and his boss cringed immediately and physically distanced themselves from me at the conference table. I mentioned that I had permission to act outside the standard corporate policy (getting three competitive bids on a specialized piece of equipment for a new product). I had run into the executive vice president at corporate headquarters several weeks earlier and he agreed that

> Casually throwing out answers to the difficult questions posed by corporate executives in high level board meetings when you are a junior engineer is not smart.

it would waste time to follow policy for this particular item. I was a little surprised when they immediately called the executive vice president at corporate to confirm my story. I was blessed that he remembered the conversation and verified the permission. I earned the respect of that corporation's senior management because I was frank about the status of new product introductions during my tenure. Unfortunately, the heartburn, headaches, and heart attacks I administered to every other management level ultimately put me on the fast track to the unemployment line. All this despite my technical brilliance and hard work!

Years later, I learned that my immediate boss had essentially the same personality as I did. The interaction of our personalities did not cause him direct problems, because, for the most part, he let me have my way when it came to accomplishing my projects. He was actually told to tone down my performance reviews. Problems arose because our department, with him at the head and me as a member, became known as impossible to work with, despite the five other qualified and definitely less obnoxious members. I was unaware of how the corporate entity was perceiving individual and departmental personalities. Big mistake.

Matters were made worse by the fact that my education level differed from that of the other departmental technical members and the corporate membership. My immediate supervisor and I had four-year engineering degrees, while almost everyone else had two-year technical degrees or had graduated from General Motors Institute (GMI). My experience level was considerably less than my peers and the managers with whom I interacted during projects.

The *Rule of Threes Matrix* was born in that turbulent soup, but a decade had to pass before I began to understand just how many interaction levels I was stumbling through with my shining sword of technical brilliance.

I essentially repeated my "bull-in-the-china-shop performance" as a middle manager at a privately held firm. The upper level management recognized my technical brilliance (possibly because I polished it often and held it up for all to see) and hoped that more responsibility and latitude would help me develop better awareness and a better attitude. Instead, as I managed two departments, I taught my people to be technically brilliant. Lucky for them, most understood my shortcomings in the political arena and didn't repeat my mistakes, especially since I was asked to leave that corporation as well.

I gained most of my business acumen after starting my own company. Clients do not tolerate obnoxious behavior from vendors except in very limited cases, and I didn't qualify for any special consideration. It was about this time that OFE&M started becoming a clear concept.

One of the core concepts of OFE&M is that one must understand corporate politics, not play them.

Effectiveness as a manager or engineer is as much about people skills, self-awareness, and the understanding of multiple interaction levels as it is about knowledge and how that knowledge is applied. Simple stuff to understand with 20/20 hindsight.

Years after I went into business for myself, I was gratified when a client retained my consulting services to address a touchy internal political situation. It seems that within this small, but growing, family business, each of three brothers had aligned themselves with a different corporate faction. Sales, accounting, and operations each had a favorite son as a leader, and most project implementation came down to the interaction of the brothers' personalities. The operations group had identified the need for new equipment and was having difficulties justifying the project to the accounting and sales groups. I successfully came in and acted as a translator to validate the project and ease the accounting department's concerns. (I speak fluent manufacturing-ese and my accountant-speak is better than my German.) In fact, it was determined that the project should have been implemented three years earlier, but lack of effective communication and interaction had muddied the waters.

At this point, some of you may be laughing at my early naiveté in corporate politics. Or, if you are also early in your career, you may be thanking me for helping you avoid some serious mistakes. One of the core concepts of OFE&M is that one must understand corporate politics, not play them. I cannot emphasize that point enough.

A vice president of manufacturing, new to his office furniture manufacturing company and his department, called me in as a consultant. The drawer slides they manufactured on-site and installed in their furniture were failing prematurely. Forty thousand cycles versus a design life of 100,000 cycles is enough to start the fingers-of-blame pointing, and that was just what the VP was discovering in his new department. He understood the realities and the perceptions that his department was ineffective. The company had recently been purchased by a German parent company. With typical German diplomacy, the "parents" pointed out that the local talent was highly incompetent. He determined it would be healthiest if he called in an unbiased outside observer (me) to uncover the problems buried in office politics quagmire.

Politics are a fact of corporate life, but if you become adept at parrying thrusts in your direction, and don't instigate games of your own, you will have more time and energy to accomplish great things. Ultimately, honest success is the best armor against the slings and arrows of outrageous office politics.

I was refused any assistance for two reasons. One, the VP didn't want my opinion colored by interacting with any of the sparring factions. Two, most members of the sparring factions were openly hostile to me as an outsider. No warm fuzzies were handed out on this job!

Remember that I mentioned I speak German? The visiting German engineers kept a constant running commentary on the proceedings in their native tongue. I really wanted to tell them how rude it is to speak in a language others can't understand, but I held my tongue. I was gaining too much insider information about what they really thought of the situation and me. Besides, their opinion of me was not bright and shiny, and it would have only worsened if my rudeness in eavesdropping had come to light.

After a few days of investigation, I called a meeting to relate my findings. I recounted improper tolerance stacks, out-of-date drawings, incorrect materials application, and the slides' weak manufacturing characteristics. Then it dawned on me: each department member was being singled out in turn for his or her part in the quality problem. By the time I was finished, the only person in the room smiling was the VP! He had found an effective and non-confrontational way (at least for him) to get his department back on track. It was obviously a classic example of us (the department) versus them (me). In their anger at an outsider, they found common ground and could address the quality situation without letting politics get in the way.

That vice president's actions displayed real management savvy. He understood his department's politics enough to use them to work in his favor. The distinction will be lost on the ladder climbers and corporate jockeys that tend to permeate business today, but I hope it is clear to the rest of you. Divert the energy devoted to backstabbing and apply it positively instead. Politics are a fact of corporate life, but if you become adept at parrying thrusts in your direction, and don't instigate games of your own, you will have more time to accomplish great things. Ultimately, honest success is the best armor against the slings and arrows of

Even if you don't play chess, I recommend that you try a few games and become familiar with "thinking out" your moves. It's a great way to learn how actions and decisions made now have long-term ripple effects that don't become apparent until it is too late to modify the action.

outrageous office politics.

As a practitioner of OFE&M, I'd like you to keep in mind that every decision you make should consider the elements of safety, quality, and productivity at the individual, department, and the corporate levels. Become alternately personal and impersonal with the decisions you make. Each perspective will help you see the ramifications of your actions. Even if you don't play chess, I recommend that you try a few games and become familiar with "thinking out" your moves. It's a great way to learn how actions and decisions made now have long-term ripple effects that don't become apparent until it is too late to modify the action. If it weren't for the fact that chess is typically played indoors and golf outdoors, I would lead a revolt to make chess "the" business recreation of choice. (I'm not sure I would have any better luck winning at chess than I do at golf, but at least my shoes wouldn't hurt my feet, and I wouldn't get lost in the woods anymore.)

> No matter how superb or technically brilliant your work is, it cannot be considered successfully implemented unless it is self-sustaining.

Another consideration to keep on a dust-free shelf in your mind is that no matter how superb or technically brilliant your work is, it cannot be considered successfully implemented unless it is self-sustaining. Self-sustaining projects are those that, when put into action, feed on themselves and continue, in at least a modified form, to achieve the goals set out for them. A handtool rusting on the floor because it is too bulky and heavy has the same impact as the new tracking system that is so complicated, it never gets used. No one understands it, let alone derives useful information from it.

The follow-up stage of project implementation will often point out mistakes in design or the method utilized. You must admit mistakes immediately and begin correcting them as soon as possible (see Chap. 2). Practitioners of OFE&M don't waste time avoiding mistake ownership, nor do they waste energy covering it up as a dirty little secret. There are enough texts on the topic of problem-solving and project management, so I won't bother going into detail on the subject. Suffice it to say that most of us know the steps: identifying the problem, gathering data, analyzing data, generating alternatives, analyzing alternatives, implementing the chosen alternative, doing follow-up and adjusting the solution as needed to meet stated goals. If you master the *Rule of Threes Matrix* at all levels, you will have a solid foundation on which solid projects may be built.

> Practitioners of OFE&M don't waste time avoiding mistake ownership, nor do they waste energy covering it up as a dirty little secret.

# CHAPTER 4

# The Old-Fashioned Manager

Use your insight and abilities
to subtly guide, steer,
and support your personnel.

DAYDREAM, IF YOU WILL, about who you would like to have as your boss. (Skip the part about maximum pay raises with minimum expectations in work output and attendance simply because they respect and like you as a subordinate.) Once you have that image firmly in your mind, realize that even if you emulate the dream, your employees may want to run you out of the company on a rail.

An OFE&M manager must first and foremost maintain awareness of the *Rule of Threes Matrix.* Get to know your own personal style and interaction comfort level and how it will mesh with your subordinates' characteristics both as individuals and in a corporate group.

My personal management style may be too friendly, intuitive, and open for most managers, but it serves me well most of the time. One of my earliest experiences as a manager came when I was teaching a group of technicians how to program and use a computerized coordinate measuring machine (CMM). This was early in the game of computerized anything, let alone a practical and complex tool that could be used on the manufacturing floor. All the technicians were competent when it came to using old measuring tools and techniques, but were intimidated by the new technology. Progress in the class was slow, but satisfactory.

I had a feeling about one of the techs and asked to speak to her alone as we were about to take a break. A few straightforward questions later, I learned that she had taken a night class on computer programming at the local community college. She quickly added that no one, at any level, had asked if any of the employees had computer experience. She felt more comfortable not volunteering the information rather than running the risk that she wouldn't be given credit for her knowledge or initiative at taking a class. I recruited her as an assistant instructor. The CMM instruction took off rapidly after the other technicians realized that one of their own was comfortable with the new technology and was willing to share her knowledge. My personal, intuitive style of management gave me an edge in that particular assignment.

*Get comfortable with proper delegation and be lavish with honest praise for employees that meet goals and support you as a manager.*

If you truly are only comfortable being an iron-handed disciplinarian who refuses to show a human side, learn to make it work for you. Consistency, fairness, and clear communication are essential if you want to prevent revolt and maintain an effective presence with your employees. I don't particularly approve of one old employer's management style, even though in its way it was consistent, unbiased, and clearly communicated. As each of his middle managers was about to take a yearly vacation, he would call that manager into his office. He then would clearly threaten to fire him or her if he or she didn't discover a way to improve work performance while on vacation. Every manager got the same speech without fail, and we all were assured that if anything bad happened in the plant one of us would take the blame, not he. Needless to say, this management style did not get the most out of the personnel and does not fall under the auspices of OFE&M.

Allowing feedback in the communications loop can make a manager much more effective. I personally believe in being openly empathetic and supportive. Get comfortable with proper delegation and be lavish with honest praise for employees that meet goals and support you as a manager. An excellent example of this occurred when I was a young die design engineer. It involved my supervisor on a blank calculation project.

Consistency, fairness, and clear communication are essential if you want to prevent revolt and maintain an effective presence with your employees.

A properly designed blank fills a forming die with a minimum amount of flash, or excess material. Too much material can cause over-tonnage of a press, damage to the die set, or merely wasted material that is trimmed off. Too little material can create bad parts or premature and uneven wear on the trim die. Total volume and shape of a blank are equally important. You must have experience to design a blank well, especially if the final part is relatively complex. Naturally, this type of design must be done under an incredibly tight time schedule or there would be no challenge (Murphy's law).

My supervisor felt that pushing the limits of my knowledge and skill was the best way to train me. Therefore, he explained the project's time frame and the critical design parameters I was to fulfill. His experience with this particular family of parts had led him to develop a fairly standardized worksheet to expedite blank design. I breathed a sigh of relief at that bit of news but immediately regretted it when he announced I could only use his worksheet to verify and check my own calculations. In other words, he was not going to give me any guidance. I would have to develop my own formulas that addressed the design problem because he wanted to observe how I solved the problem. It was understood that his worksheet was a safety valve if I was unsuccessful.

A day and a half later, with half a dozen pages of calculations in hand, I asked for his design worksheet. Using his formulas, I was able to design a blank in about four hours. When we compared the two designs it was readily apparent that they were almost identical down a ten-thousandth of an inch in most characteristics. Some of the formulas were similar and some were completely different, but we had arrived at an overall blank design that was, for all practical purposes, identical. The significance of this eluded me even after he made it clear how pleased he was with my efforts. It wasn't until other members of the department started complimenting me (he also praised me to everyone else) that I realized I had successfully addressed a difficult design problem. Despite a tight schedule and my relative lack of experience, my supervisor had given me free rein and, consequently, I gained valuable experience on several levels.

An OFE&M manager never bullies or rides shotgun on an employee or project team. Those people who have true power never let it show directly. Use your insight and abilities to subtly guide, steer, and support your personnel. The more initiative you allow, the greater the accomplishments you will get to oversee. Channeling and focusing creative energy are essential. Find your own personal way of doing these things without being controlling and stifling.

A simple self-check is to view your department as a mirror of your management style and effectiveness. This doesn't mean you take credit for all successes and blame for all failures. (Worse yet, don't take credit and sidestep blame.) It simply means that, if your department is highly regarded, then you are being effective. If your department is being eyed askance, review your characteristics as a manager and see how the *Rule of Threes Matrix* interactions are stacking up. Perhaps a change is in order.

One of my martial arts instructors loved going to tournaments and observing his students compete. He told me he learned more about his skills as an instructor and as a martial artist watching the techniques his students were using than by any other method. Many times what he was teaching and what was being learned were two very different things. He took full responsibility if any student displayed an unsportsmanlike attitude or bad technique. I eventually learned to recognize how he was going to modify our lessons by watching my fellow students. Most competitive teams demonstrate their coaches' characteristics, and as a manager, one of your roles is as a coach. If you don't get your monthly department report in on time, don't expect your people to get their monthly project reports in on time either.

> Those people who have true power never let it show directly. Use your insight and abilities to subtly guide, steer, and support your personnel. The more initiative you allow, the greater the accomplishments you will get to oversee.

An unfortunate example involved an engineering manager and his department at a company experiencing a severe ergonomics problem. The manager's response to most complaints was to dance around the facts so that he avoided expending any effort in actual project work. He was so talented at this, he even took on the EPA and OSHA at the "paperwork shuffle game" and won. While he was dancing the dance, the problems the operators were having did not go away, but got worse. When a toxic fume level was technically below NIOSH limits, the manager refused to address the issue. Never mind that many operators and supervisors were physically ill from long-term exposure, this manager loved playing games and would not budge from his position.

His department ultimately operated in the same fashion. They were manufacturing engineers. They never went out on the floor, nor did not they know any operators by name. A quick-adjust table they designed weighed in at more than 240 pounds, but they never saw it on the floor and didn't care if it was effective or not. This same department designed custom equipment that used SAE hardware, even though they knew the machine mechanics weren't allowed to have SAE wrenches around the metric-designed machinery.

Interestingly, this engineering department was very good at playing computer golf on their very expensive CADD workstations. I think the engineering manager had the lowest handicap in the entire department.

# CHAPTER 5

# The Old-Fashioned Engineer

The human characteristics
in engineering should be considered
from two different viewpoints and should
never be left to chance.

TAKE ALL THE NEGATIVE CLICHÉS and bad jokes about the typical engineer and try to be the exact opposite. Our problem-solving training sets us engineers up for some bad traits when it comes to dealing with people and organizations. Yet, oddly enough, when we apply the same problem-solving and intuitive skills to working with people and organizations, the results, more often than not, are gratifying. Engineers just need to realize that it is a valid exercise and not a wasted effort.

Beware: no matter how good your formal engineering education is, the university path and the degree-by-experience both come up short, leaving holes in the practical end of things. Most engineers wield the sword of technical brilliance with ease and some even exhibit a certain flair. How could a techno-phobe possibly be an engineer? However, the human and practical variables most practicing engineers leave to "work themselves out" comprise the balance of the successful OFE&M engineering equation.

The human characteristics in engineering should be considered from two different viewpoints and should never be left to chance. These viewpoints are that of the design's end user and that of the manager supervising the engineer.

Any design an OFE&M engineer undertakes should constantly be checked from the operator's perspective, even if that means the operator works hand-in-hand with you in the development process. This will speed the process and eliminate a lot of backtracking or, in the worst case scenario, a product that dies because the customer can't use it. By working with the end user, engineers generally are forced to confront whether the design is simple, inexpensive, safe, flexible, and clean and is still effective. Inexpensive means cost justifiable and effective generally means that quality and productivity issues are successfully addressed.

Several examples that illustrate this facet of OFE&M jump to mind. Those quick-adjust tables that weighed 240 pounds and took three operators and special tools to change the height are a classic. The engineer in question did not even know the operators' names. When faced with the reality that his design could not be adjusted in the fifteen seconds allocated, he literally ran off the production floor and into the office. I ended up working with the line workers. We generated ideas that we presented to engineering as being livable. The in-house engineering staff never did learn my proper name, but they did learn to resent my designs and the fact that I got along well with the operators.

I also worked with a handful of CADD engineers who

> I don't care how good you or your CADD system is, predicting all the variables and how they will interact when someone else is trying to mate interrelated parts is difficult.

had good formal educations; however, they had never worked with fabrication equipment. They had no clue how the phone booth housings they designed actually went together on the assembly line. The press and brake operators were usually in trouble with assembly operators. The assembly operators blamed the press and brake operators for bad parts that actually could almost always be traced back to simple engineering design errors. My first question was why the engineers didn't spend more time learning firsthand how a CADD design became a real product. It was a foreign concept to them at first, but they quickly realized the value of leaving the CRT and "playing" on the floor. I don't care how good you or your CADD system is, predicting all the variables and how they will interact when someone else is trying to mate interrelated parts is difficult. Tolerance stacks mean one thing on a calculator. They mean quite another when a security bolt is being threaded into a weld nut on the spot where four sheet metal plates with various materials, thicknesses, and configurations intersect.

If, as an engineer, you can truly learn to listen to line worker input, no matter how irrelevant it seems at first, your job will become infinitely easier.

I have had to learn not to let my conscience bother me about working with production employees on my workstation or product designs. If, as an engineer, you can truly learn to listen to line worker input, no matter how irrelevant it seems at first, your job will become infinitely easier. I have had designs laid in my lap that needed nothing more than to be rendered into fabrication drawings because I asked an employee who had long been chafing under an old design for his or her opinion. Realize also that not giving someone proper credit for the design will cause this well of information to dry up rather quickly. Instead, give credit to the

employee for the design and take credit for asking. I have
even gone as far as allowing said employee to make a color

selection on any additional equipment or fixtures as a thank you. It
may seem trivial or ridiculous the
first time you try it, but the proof is

Give credit to the employee for the
design and take credit for asking.

in the pudding, and you will love this recipe.

Documentation is also critical to the OFE&M engineer.
I have learned to write down the date, time, and personnel
involved with a particular project at the top of every page I
use. To be able to trace time and design lines months or
years later by following the dates of the pages in a given file
is a great time saver. This habit has often yielded quick
answers to simple questions that couldn't be answered without it. For instance, a VP may want know why a design
change in a production layout was made. You may not
remember why offhand, but if you have the notes and dates
of all the related meetings, looking up the requested information is a simple matter. (This is, of course, assuming that
you weren't sleeping with your eyes open during those
marathon board meetings.)

Proper documentation saves time because it prevents
redundancy and the recurrence of previous mistakes. This
leads to giving proper consideration to the manager overseeing a given project. Engineers tend to focus on one aspect of
a project at a time, even when they're aware of others. These
are discounted until they appear on the agenda, to the detriment of supervising managers held responsible for overall
time tables and results. Many, if not most, managers will give
engineers more leeway to "engineer" as they see fit if the
engineer gives them at least a modicum of consideration.
Timely engineering reports and quick notification of problems (as well as positives) made without prompting go a long
way. I won the eternal gratitude of a client who managed a

fiberglass fabrication facility by making a timely, clear, and concise engineering report that described the operations currently performed on the floor. He was no longer in step with the changes occurring there. As a result, the project he asked me to investigate was no longer in the critical path. I could have spent a lot of time and money completing the assignment, but I saved the client embarrassment and money instead. I don't know if he was more surprised at the changes or the fact that, as a consultant, I had kept him from wasting money I could have put in my pocket.

Documentation is also critical to the OFE&M engineer.

Always keep in mind the *Rule of Threes Matrix* and the interactions swirling around you as you display technical brilliance. Make sure you are lighting the way to technical dominance for the department and the corporation, not simply giving everyone a headache because of the glare you are indiscriminately radiating.

Use what you already know, even if it doesn't fit neatly into the technical brilliance category. Many of us love to play with our techno toys and refuse to participate if techno toys are not available. I've even known engineers that collect technical techniques and equipment on the off chance they may need it later, when basic information they already possess could successfully address a given situation. A colleague, a quality assurance engineer in a greeting card/novelty gift plant, recently gave me a good example.

As the story goes, the candle production department was having a terrible quality problem. Getting good parts consistently was difficult. "Jill" (names have been changed to protect the competent), a relatively young engineer, was told that the only way to address the problem was by changing the wax formulation to compensate for whatever was causing the rejected parts. It seems that any time the department encountered a production problem, the knee-jerk reaction was to play with the wax formulations until some parts came out right.

Wax formulations are, or can be, very sensitive to many variables. Jill made the mistake of asking if all variables had been investigated thoroughly before something as major as changing base formulations was undertaken. Mold design and cooling jackets, temperatures, humidity, equipment maintenance, and a host of other factors all had been ignored. Actually, they had not even been documented or tracked before. Jill laughed as she related her quality success story over lunch. She thought it was funny that, although she did not have any previous candlemaking training, she was able to address the problem with timeliness by relating candle manufacturing to something she did know. The basic knowledge base she drew from? "I just figured that it didn't seem much different from baking. I follow a set recipe in baking cookies, why not use one in making candles?"

> Technical brilliance has its place, but common sense rates up there with the million-dollar solutions every day.

You might be surprised how prevalent this kind of situation is in manufacturing and business today. Technical brilliance has its place, but common sense rates up there with the million-dollar solutions every day. One of my first assignments after college was to fly halfway across the country and oversee the loading of used woodworking equipment being transferred to our Arkansas plant. I didn't know V-grooving equipment from a nuclear reactor, but I did as I was told. When we got the equipment set up in the plant, the V-groover was a beast possessed. It refused to do anything right and seemed intent on beating itself to death. I was not in charge of setting up the equipment, but as I had overseen its loading, it was assumed I would know how the problem could be remedied.

I was behind the eight ball in a big way. I had been on the job less than a month, and the equipment (even used) had to be functional as soon as possible. I had very little experience with automation and none with hydraulic logic. Instead of panicking, I stared at the equipment and tried to envision how it worked and how the designed-in operations could be completed logically. I walked through the operations as if I were doing them by hand and then studied the machinery components to decide how it would approach the same procedures. It took me almost two hours of "sitting on my butt" and staring in amazement at that conglomeration of hoses, wires, and fabricated parts, but I figured out the problem. Some hoses had been reversed on the hydraulic pump, and the V-groover was trying to run backwards! Naturally, parts were crashing into one another, because they were in the path of parts that had yet to function and clear. After the hoses were reversed and the system was bled, it operated perfectly. Never underestimate the value of sitting and thinking.

*Never underestimate the value of sitting and thinking.*

Years later, I was faced with an equally perplexing problem that seemed simple enough, yet it presented no clear and simple solutions. In a plant where I oversaw the engineering and maintenance functions, we had several automatic buffing machines that used a vacuum to hold pieces of cookware on chucks. Twenty-five-horsepower motors with attached pads buffed the cookware. Unfortunately, the vacuum that held the cookware in place worked so well that it also sucked buffing compound and buff material into the its lines and eventually into its pump. Once in the pump, it was only a matter of time before the compound did its polishing magic on the pump bearings, and we experienced pump failure. This happened, on the average, about once a month and cost approximately $1500 in parts as well as my maintenance tech's time and lost production costs. (Not to mention which, when the vacuum failed, that work area became hazardous because the cookware would fly around the room and bounce off the walls.) The bills were mounting up, and we needed a solution. In-line filters had been tried, but they clogged so easily that they also impeded production. Master bag filters at the pump were in place, but they filled up so fast that keeping them clean and preventing their failure was another production nightmare.

The solution in the end was simple enough and was based on common sense, tempered by parallel thinking and a touch of school-book learning. My maintenance department was not opposed to sitting around and thinking out loud when they faced a problem. (Perhaps it was the fact that I didn't treat it as a crime, but encouraged it, and occasionally even joined them.) I believe the fact that we generated viable solutions elevated these to "brainstorming sessions" from the lowly "bull" sessions other departments employed. Anyway, one tech pointed out that in the HVAC industry he had seen tanks used as in-line filters, although he wasn't sure of the internal configuration. His observation led us to determine that we needed a "drop out" tank to catch the compound and buff material suspended in the air stream fed to the vacuum pump. I knew from fluids class in school that, if the inlet of the tank was smaller than the outlet, we could design a velocity drop in the stream and possibly slow down the flow to a point where it wouldn't support the contaminants. Because the inlet and outlet had to be plumbed to standard piping, we made a steel box with internal plates to simulate progressively larger diameter piping. We also designed it to have a turbulent flow so that, after the velocity dropped, the stream went up and over a final plate and the likelihood of the contaminants losing energy and dropping to the bottom was increased. The internal plates also supported the sides of our box to keep it from collapsing from the vacuum. Knowing full well who would have to clean it out, we placed flush plugs on the top and bottom and welded eye attachments on top so a hoist could be used to place the whole contraption into our degreaser.

We only spent a week and $500 on the project and really didn't have tremendously high hopes for success, but we knew we had to try. To make a long story short, the dropout tank worked so well that the bag filters didn't have to be changed as frequently, and they didn't fail so often. We determined that degreasing the tank once a quarter was sufficient to maintain reliability. We didn't even lose any production because a designed-in bypass (another tech's idea) allowed the bag filter to do the job by itself while the tank was flushed. Not bad for an informal group discussion that addressed our woes. Common sense, experience, and communication solved a lot of problems in that facility. Just about anyone could have questioned our technical brilliance and our lack of big budget dollars was no secret, yet we still got the job done.

Many times it is just such a simple approach combined with horse sense that will see an engineer through tough and perplexing situations.

# CHAPTER 6

# Cliches

Cliche's are simple phrases.
They remind us that we all have common sense
and knowledge but we tend to ignore them.
It is for this reason that I like to use clichés
to relate the higher concepts of OFE&M.

## KISS — Keep It Simple, Stupid

I HAVE HEARD THIS SEVERAL HUNDRED TIMES in my career—the most embarrassing being when I was muttering it under my breath to myself. The advent of the pocket calculator (later the programmable pocket calculator and ultimately the giga-byte laptop computer with active matrix color monitor and built-in printer) redefined engineering design and management decision-making. Nowadays, it is quite easy to crunch massive quantities of numbers quickly. In his article, "How Engineers Lose Touch" (American Heritage of Invention and Technology, Winter 1993), Eugene S. Ferguson takes a thought-provoking and sometimes sobering look at modern engineering practices. Just because we now have the technology to create exceedingly complex designs does not necessarily mean that such designs will inherently be better than a simple one.

There are many more opportunities for making mistakes in judgment and even in simple calculations with complex designs. The more convoluted the design, the greater the risk of catastrophic failure and the higher the cost of initial implementation and system maintenance. I have heard it said that as a system's complexity approaches infinity, the mean time to failure approaches zero. I haven't seen any mathematical proof that this is true, but anyone I know who has had to maintain equipment won't argue that a simple mechanism (or circuit) is much less likely to fail. Complex organizational structures and command chains operate under the same set of rules. Also, there is an inherent elegance in a simple design that addresses a complex problem.

Rather than using a classic example of bridge or dome design and failure, I'll relate a simple and personal example. The new(er) car I own has a marvelous device on it called a "Heads Up Display" (HUD) that projects my speed onto the windshield. The analog signal that tracks my car's forward progress is encoded into a digital signal that lights up LEDs, which are then projected onto a special section of treated glass. I can then see it in the left-hand corner of my windshield, and I don't have to look down at my dash. I was thrilled with the new toy (as the marketing people hoped I would be) and I delighted in showing it off. I never really thought about it from an engineering design viewpoint until it failed. Even then, I really didn't give it much thought until I saw the estimate for repairing/replacing the unit, and it was more than $400.

> The more convoluted the design, the greater the risk of catastrophic failure and the higher the cost of initial implementation and system maintenance.

My old 1967 Mustang used a mechanical cable, magnet, and indicator needle arrangement that was incredibly simple in design and function. It was also incredibly inexpensive to fix—I repaired it once with $15 in parts and half an hour of labor. The service manager at the new car dealership told me that the electronic probes on the test meters that could fix my new speedometer are hair fine. It takes more time to learn how to use the computer diagnostics than it took to repair my old speedometer. Both designs record speed with about the same accuracy (as far as the police are concerned), but I really missed my old Mustang as I wrote out the check for the repair bill.

*Just because we now have the technology to create exceedingly complex designs does not necessarily mean that such designs will inherently be better than a simple one.*

One of the most positive examples of the KISS principle I know was used by a "seat of the pants" design engineer known as K.A. (don't forget, even the initials have been changed to protect the brilliant). K.A. needed several hundred hold-down clamps that could be hydraulically or pneumatically operated for production welding tables. This is not a problem if you have a few hundred thousand dollars just for parts in your budget. The size, clamping pressure (4500 lb.), and remote operation characteristics eliminated any commercial clamping systems due to cost. K. A., a resourceful fellow and part-time race-car driver and builder, stepped back and took another look at the application.

His simple solution was effective and saved both time and money. He purchased truck brake components based on the fact that brakes act as simple clamps. I can attest to the fact that the welding tables work very well. The controller at his plant can attest to how inexpensively they were built.

Of course, for every positive example of the KISS principle in action, there are hundreds of fiascos implemented in the name of good engineering or management. The cream of the crop, the absolute hands-down winner in my career involved a new line of commercial, thermally broken aluminum replacement windows. In a thermally broken commercial window, the individual glass lite frames typically ride on a thermally non-conductive guide attached to the sides of the larger window frame. The guide separates the metal components and precludes heat transfer in either direction. Expensive and high-tech, "quality" was the buzz word for these windows. No design or manufacturing aspect went unnoticed as the project progressed. Every detail was designed and redesigned to perfection.

As the local manufacturing engineer, I had the dubious distinction of having to handle both the corporate product designers and the sales representatives whose only baby was this new product. The pressure to produce product quickly and effectively was enormous. I had been on the project about seven months and could quote part numbers and assembly components from memory. I could even sketch cut-away views by rote and describe fabrication procedures in my sleep. We had extruded several sticks of prototype components, and we had begun fit-up tests that appeared to be progressing quite well. Therefore, it was with some consternation that I overheard the corporate "big dogs" worriedly whispering about a design flaw they had discovered at this late date, just prior to introduction.

The designers had come up with an ingenious idea to make assembly easier. They had an arrow extruded with two barbs on the metal window frame's inner side. Then a special plastic guide was extruded with a negative image that matched the arrow so that it could be slipped over the frame component before assembly. After the frame was assembled, the head and sill locked the guide in place. Because they were unsure of the clearances necessary for the guide to slide into place, the designers allowed a quarter inch of clearance along the arrow's shaft, with only the "head" of the plastic guide and aluminum arrow locking together. The line workers and I were actually quite pleased with the easy assembly and had not detected any design flaws. They worked quite well, in fact.

In clear and unflattering terms, the designers pointed out our ignorance. It seems that the designers, after seeing a cutaway cross section of the two assembled components were horrified at the aesthetics. According to them, two barbs on the aluminum shaft didn't have corresponding cut-outs in the plastic guide (a major sin). The tolerances were so sloppy, they claimed, that the barbs weren't needed to slide the guide in place! The designers wandered off muttering. I was understandably confused because I knew no one would ever see that particular cross-section except in a salesman's cutaway sample. The incident was forgotten until two weeks later when the designers returned with a new and aesthetically pleasing guide cross-section.

The new design used close metal tolerances of thousandths of an inch and was indeed very pretty. Unfortunately, after pounding on the end of a three-foot plastic extruded section, no one could coax it into sliding over the mating aluminum arrow. It seems the tight toleranced mating surfaces created too much friction. Even good old WD-40 and tons of cussing could not get that part to slide on. Even in that pre-PC and simulation world, the workers predicted trouble for the entire line. Some of the larger window guides were almost seven feet long and would be even more difficult to assemble.

As a system's complexity approaches infinity, the mean time to failure approaches zero.

A few feet of the older, simpler guides were rounded up

quickly to complete the assembly tests, and the old die was ultimately used for production. At the time, watching the temporary stand-off between the designers and the assembly workers was fascinating: one side was armed with cans of WD-40 and dead blow hammers and the other had colorful cross-section CADD printouts and righteous indignation. Looking back, I see a certain irony in using soft-faced dead blow hammers; at one point they almost lived up to their monikers.

To this day, "cross-sectional aesthetics" is a phrase that brings a smile to my face. It has actually caused some embarrassment over the years. An engineer in that department came to visit my family (he was introducing his fiancee to us) years later, and he made the mistake of asking, "Remember cross-sectional aesthetics?" Everyone in the room thought we had lost our minds as we laughed until we cried at the memory of those white-shirted designers up to their elbows in lubricant and desperately trying to force those pretty cross sections onto the frame members. KISS.

## There Is Nothing New Under the Sun

I AM NOT COMPLETELY SURE IF THE REST OF THE WORLD has the same propensity for the generation of descriptive verbiage in relationship to fundamental conceptualizations as the United States of America, but I sincerely hope not. (Translation: I hope that the rest of the world doesn't create lots of new words for the same old ideas as we do in the United States.)

Perhaps it is political correctness or perhaps it is the fact that we have more lawyers per capita than anywhere else in the world that causes us to repackage the same old goods in new wrappers. Personnel departments have been replaced with "human resources" departments or "employee involvement centers." The people working in these offices do not look any different, nor do their offices and office equipment. Of course, I've been told that I just don't fully appreciate the subtlety of the titles.

Secretaries are now "executive assistants," "personal time managers" or "office managers." I can't think of any manufacturing facility that employs plain engineers any more. I have known engineers whose duties did not change, but

The basic facets of management do not change except as seminar titles for the upcoming season.

their titles rotated through production, manufacturing, industrial, process, project, product, resident, facility, and design engineer.

Usually, the only people who note the changes with any concern are the draftspersons (design layout technicians, CAD operators, rendering specialists) who make the necessary changes to organizational charts. However, there are occasional exceptions. For example, have you ever known a manager or engineer who was brilliant, yet got so caught up in his or her title or position that it appeared he or she had forgotten everything else? A friend of mine who had managed a large glass shop successfully for many years decided to start his own glass company. Within two years he was almost bankrupt and ready to throw in the towel. I sat down with him and started analyzing the books, looking at margins and cost centers. About halfway through our meeting, I started asking generic questions about how the glass business works, hoping to get a feel for it myself. After a few hours of discussion, it became apparent that the only real problem with the business was that my friend had become enamored with his role and title. Because of this, he had not used the basic work and managerial habits that had made him a successful manager when he worked for someone else.

I am happy to say that the business is growing, and he has opened his second shop. I didn't tell him anything new about how to run his business; I just reminded him of the knowledge he already possessed. I feel pretty good about that job, because I charged him nothing and helped him save his business. Of course, the accounting/managerial consulting firm he hired before I looked at his books charged him $10,000 and didn't even do that. They did manage to confirm that his expenses were much larger than his income. They recommended that he do something to increase revenues and lower his out-of-pocket expenses. I suggested that not paying the $10,000 would be an excellent start, but he paid the bill anyway.

Oddly enough, my friend and his glass shops are still in business, but the consulting firm is not. Go figure.

Management techniques and styles are as susceptible to these name changes as personnel titles. The basic facets of management do not change except as seminar titles for the upcoming season. My current favorite is "reengineering." I recently received a brochure in the mail that describe a dozen seminars. The brochure was almost identical to one I received the previous year except that ". . . in a re-engineering environment" had been tacked to the end of most titles. I have never been one to jump onto the bandwagon until I know where it is headed, and in the case of reengineering, I think I have seen this road before from the same seat.

Technology can make us think we are seeing something new because it is faster, brighter, neater, or more expensive. Drafting is a wonderful example because the end results (drawings) are identical when produced by a skilled draftsperson in either media. Technology can also give you a false sense of expertise, to the point of embarrassment.

A potential client was using a popular CAD package and wanted to make sure that the CAD package I use was compatible and that files could be exchanged easily. They downloaded a drawing from their system to disk, and I took it home to see if I could modify it on my system. I opened the file to discover that the drawing was 508 inches by 762 inches. Other than that little anomaly, I had no trouble making modifications. But, when we reopened the file on the client's system, the drawing had been reduced to minuscule proportions. After a great deal of investigation, it was determined that because the designers preferred working in inches and the product was made to metric specifications, they simply worked in inches and then scaled the drawing by 25.4 to "have it rendered in millimeters." In other words, they were making the drawings 25 times bigger than life!

*If you aren't doing it right without the program, then adding a computer and really good software to the mix simply means you are still accomplishing very little, but are doing so in a hi-tech manner.*

The designers involved had never worked on old-fashioned drawing boards. They had never had to lay out a series of views to assure that all would fit on a fixed size of paper. Nor had they ever scaled a drawing by hand. In addition, they had no experience with the metric system, and they were self-taught on the computer. They admitted that printing out drawings and exchanging files with overseas divisions had created scaling inconsistencies. They never did admit to understanding why scaling one inch by 25.4 did not yield millimeters.

Board-experienced draftspersons will laugh because scaling is another term for multiplication, as the programmers/designers for CAD software know. Just because the computer makes generation and modification of drawings quick and easy, that does not mean the basic concepts have changed.

Time and time again, I have heard the horror stories. They all boil down to this: if you aren't doing it right without the program, then adding a computer and really good software to the mix simply means you are still accomplishing very little, but are doing it in a hi-tech manner. For instance, preventive maintenance (PM) programs can be administered quite simply with a notebook and a few basic NCR (no-carbon-required) forms. I have set up PM programs for a relatively large, true just-in-time (JIT) facility this way. The system's simplicity is what allows it to work. Granted, automatically generating work orders and tracking parts are nice perks, but the core of any PM program is having good facts on which to base decisions. This is a universal concept. Once the basic information is gathered and verified, that is the time to purchase the software and enter the data into the system. If you know it works with a pencil and paper, then purchase the computer and use the technology to make a good system work faster and better. That is its purpose. Expecting technology to be a magic wand is just setting yourself up for an expensive disappointment.

Another classic example of technology and terminology not changing the basics comes from a friend of mine, a manager of a large corporation's telemarketing department, who called me for some ergonomics design help. She was new to the company. She had been given two weeks to prepare a budget for completely redesigning her department—from ceiling tiles to carpeting and everything in-between. The project should have been relatively simple. Unfortunately, her company (as many large corporations tend to do) had made an agreement with an office equipment supplier for volume discounts.

This in and of itself was not a concern until it was time to

*Expecting technology to be a magic wand is just setting yourself up for an expensive disappointment.*

select a suitable chair design. The "ergo-nomic" chairs the supplier offered were a classic example of a widely recognized manufacturer being slow to respond to market demands for a better chair. The chairs in question were beautifully sculpted, came in design- er colors, were manufactured with the latest technology, and were incredibly expensive. They did not, however, have a full range of adjustments, nor were they particularly comfortable. Because the basic height was adjustable (very few office chairs made today do not have this feature), the manufacturer deemed it an "ergonomic" design.

"There is nothing new under the sun" also has a rider: "Don't reinvent the wheel."

The equipment company's representative responded somewhat negatively to my requests for a more adjustable chair. In fact, she rather pointedly asked me, "Who the f#@* do you think you are?" I briefly explained my duties and responsibilities to my client. I added that I could not rec- ommend an expensive and unsuitable chair just because the company had used the proper technical term when they named the product. After a few more comments about my lineage, she finally admitted that the chair design was not very good, and that the company would not have a fully adjustable chair ready for about six months. Since our dead- line was two months, a special waiver to the purchasing con- tract was allowed, and another company supplied chairs.

"There is nothing new under the sun" also has a rider: "Don't reinvent the wheel."

You can use the fact that there is nothing new under the sun to your advantage and prevent needless time, money, and energy expenditures. I have a friend who is a research scien- tist and believes in the basics (i.e., OFE&M, if you will indulge me here). He was hired by a huge and world- renowned corporation to research collagen eye shields. The corporation hoped his research would lead to innovative breakthroughs and ultimately some patented products. His first question upon entering the lab was, "Where is the literature search information on related products, research, and patents?" He was quickly informed that he was not hired to do literature searches or even to review literature searches. Fred (of course, this is a fictitious name) had been taught that a good scientist always reviews published material so that he has a solid base upon which to do innovative research.

Ignoring his bosses, Fred did a literature search anyway (on his own time), and guess what? He discovered that someone in a different field had done a very large body of collagen work. The base material manufacturing and processing techniques were almost the same. This fact meant he could research the specific application areas, and it ultimately led him to several patentable products within a couple of years. If the corporate higher-ups had had their way, he would still be trying to establish how the base material was processed so that it could be researched. They came dangerously close to the "we-are-too-good-to-do-that-type-of-menial-task" attitude prevalent in many corporate structures today.

## Do It Right the First Time

As a boy, my teachers often admonished me to use foresight and plan my term papers and class projects. Then, I would not have to redo the work, and I would ultimately be happier with the overall outcome. As most young students do, I tended to ignore this advice until many years later when I discovered the wisdom of it for myself.

When I was sixteen, my father, still considered the consummate teacher by many, tried again to teach me this lesson as I was learning to rebuild/restore the 1967 Mustang that would be my transportation for the next ten years. The main difference between a term paper and a car, I soon invented, was that, compared to pencil and paper, car parts are very expensive. After a few mistakes, I realized that a few moments spent planning what, how, why, and in what order, would save me money and many frustrating rework hours. Possibly even more important, I realized that the time it took to redo the work could have been spent driving the car with my friends. Even though we usually drove in circles, the time I lost with my friends was very real and personal to me.

The corollary to "do it right the first time" might well be the rhetorical question "Why do we have the time to redo this project if we didn't have time to do it right the first time?"

I have to believe that most of us have similar backgrounds (where else do clichés come from?). So I am somewhat perplexed how, when we become managers and engineers, we forget the childhood lesson of "do it right the first time" as soon as a project rears its head and hisses at us.

One of the best lessons in project and time management I ever learned came when I was an intern tool designer at a large corporation. I was told that my student status would only buy me minimal consideration when deadlines threatened and designs were scrutinized for functionality. My mentor explained that he felt the only sensible way to remain on track with projects was to plan the work methodically. Consideration had to be given to as many variables as possible, and the schedule created must be followed. ("Plan the work & work the plan" has become a pseudo-cliché since that time and "robust design" is a really nice catch phrase currently in vogue.) Exercising good judgment at the outset of a project meant considering contingencies, and balancing them against the fact that the schedule for any important project would be moved up the closer it came to completion. The hardest part was staying focused on the plan despite the pressures to finish portions early or out of sequence. (I know a company that updates its project implementation charts weekly and the deadlines for completing any individual projects are moved almost that often.)

The only criticism of my mentor I ever heard was that, although he was a talented individual, he tended to be too detailed-oriented, and sometimes he took too long to complete projects. That criticism was then followed by the statement that no one ever had to go back and review his work for accuracy and that inevitably his projects always worked the first time.

The corollary to "do it right the first time" might well be the rhetorical question "Why do we have the time to redo this project if we didn't have time to do it right the first time?" I believe this question is invariably ignored when sales managers want prototypes for show-and-tell or when a new market for an existing product is opening. I have also observed this phenomenon around the time of year when reviews are due. I have yet to meet the manager or engineer who likes to admit in a salary review session that "the project" for the past year is not yet completed. It is much easier and usually more cost effective personally to rush a project toward completion before the review and then repair it later. Sometimes this is done in the name of "reengineering."

In the real world, I have seen many examples of why reengineering has come into play as the seminar/book/philosophy/technique of the moment. I must confess to being more intrigued by how much reengineering could have been avoided if it had been done "right the first time" and how to prevent reengineering in the future.

I know where there are fifteen, high-dollar robots (six digits each) in mothballs because a design team was too lax in its data gathering efforts to determine that, although the robots were ideal for performing the necessary functions, they were two feet too tall to fit into the space allocated. It turned out to be less expensive to buy a different brand of robot than to modify the production line. Now, every time this corporation takes on a new project, this kicks it off: "Is there any possible way that we can use fifteen nine foot tall robots in this application?" (About the only logical answer I heard was to donate the robots to a Vo-tech school and write off the cost as a charitable contribution.)

I used to be in awe of the automotive and aerospace industries until I started working in them. Big disappointment. I assumed that, with all the money and resources available for manufacturing "big-ticket" items, somehow a magical transformation allowed engineering and managerial miracles to take place.

I was asked to look into an aluminum fabrication area in an aerospace plant with ergonomics problems. The finishing process on some lightweight panels involved using old-fashioned air riveters as peening hammers. The vibration was seriously damaging soft tissue. The company was trying to find a handle style and riveter weight that would minimize the trauma. My initial reaction was to ask why the forming dies weren't retooled to eliminate the hand work. The response was that the die engineers were afraid to dial-in the presses because they were old and they didn't want to "crash" or damage them.

Instead of designing the dies so they worked in the press and then training the set-up operators to produce good quality parts, the staff decided that controlled hand rework was the answer to their problems. The cost in human terms, as well as quality and productivity, is extremely high. I am sure at some point the process will be "reengineered" to eliminate the costly handwork operations. The reason I can make that statement is very simple. I managed a plant with similar ancient cast iron presses and we used strain gauges and intensive training to address the issues. Trust me, $15,000 per press in a one-time shot goes a lot further toward the bottom line than wages and benefits for four employees each year. That doesn't even consider the health costs involved in the inevitable cumulative trauma disorder (CTD) cases.

You don't need a huge budget or a large staff to think through a job design and do it right the first time. In fact, some engineers and managers who have these luxuries still take the "quick and dirty" way out regardless of the resources available. Develop a plan and work diligently and you will be able to take advantage of the resources available to you.

## Haste Makes Waste

HASTE MAY MAKE WASTE, BUT DRAGGING YOUR FEET will lose you market share. How do you strike a balance between the two ideals without going crazy? Initially, managers or engineers must be honest with themselves about how much time it really takes to complete a project. The second step is to admit how much time can be cut from that figure when a steady, productive pace is maintained.

In many of today's organizations, more hours are spent discussing the work that must be done than in actual productive activities. This reality is viewed as an insider's joke or a nasty part of scheduling. Others see it as a safety valve in case a really important project blows up and must be finished quickly to save an account (or one's job). The viewpoint depends on your position, the corporate culture, and whether or not you are ultimately responsible for the project's successful completion.

Haste and execution speed are not interchangeable. Haste implies poor planning or panic management. In one company I worked for, there was a certain project engineer who was notorious for being impossible to find. The standard joke was that he had a bathroom stall with a trick door that allowed him to disappear at will. Amazingly, he always reappeared when corporate officers arrived at the plant door and he invariably made a good showing of his projects. No one fully understood how he managed to do this.

An old college roommate of mine (who actually took longer to get his bachelor's degree than I did) landed a traditional sales position selling non-food products to grocery stores. He had a route to follow and a quota to meet for each month. The trick, he told me, was to qualify for the top sales award the company offered, without going so far over the quota that it would be raised the following year. He collected bonuses and a regular paycheck, drove a company car, and sold sample products for cash to supplement his income. He and his wife also took a nice trip to the Caribbean at the company's expense because he was such a hard-working and responsible employee. And yet he only worked, on average, two-and-a-half days a week!

I like to believe that the company realized his work habits were below par, because after he had worked there for two years, he was terminated in the name of "downsizing." In all probability, he just simply did not have enough years of seniority when the axe fell. After all, another sales person had trained him and explained how to work the system.

In the above examples, each individual had a planned routine and work habits that defined how much work they produced. If necessary, they could produce at a far greater level than normal at a moment's notice. However, at that point, they were operating outside their comfort zones, creating inefficiency and wasted efforts. In these cases, personal work habits caused the haste, but many times haste is produced by outside factors.

Clients, upper management, or other departments applying pressure for work not yet scheduled can cause haste. When this occurs it is very difficult, if not impossible, to retain high quality standards. Often, a turbo-boosted project will fall apart shortly after completion. This reflects the inefficiencies that time crunches inevitably cause. Hindsight may come in handy at that point.

It is a rare client/boss that prefers shoddy work done quickly over good work only nominally ahead of schedule. Let's face it, we all want excellent work done fifty percent faster than estimated. However, most of us will settle for slightly faster and only slightly lower quality if a choice must be made.

## A Stitch in Time Saves Nine

THIS CLICHÉ HAS A MODERN EQUIVALENT in the adjective "proactive." Proactive is often used to modify engineering's, design's, maintenance's and management's stance on any given issue. In fact, I have heard it used as a stand-alone statement: "Is it proactive?"

I am reminded of comedian George Carlin's routine about words and their meanings whenever I hear proactive. Too many times, I think that it has become a catch phrase with no meaning, but it impresses listeners when it is thrown into a conversation. Although "proactive" is much easier to say than "a stitch in time saves nine," it does not provide a useful image. Identifying a potential problem or problem area is difficult. It is almost always accomplished more easily when 20/20 hindsight is involved. Experience is critical and is usually gained at great personal and professional cost. If someone can tell me where to place the stitch, it is usually a simple matter to do so. I believe so strongly in that single stitch that I always give the manager or engineer who has a project unravel around their ankles the benefit of the doubt. I assume that he or she merely had not realized the stitch was needed. Unfortunately, I have seen many others ignore initial warning signs or simply wait for a catastrophe so they can be heroes. In some cases, prevention can begin before there are any signs of failure (I believe the current buzz word is "predictive"), yet is often ignored because of time or budget constraints.

Obvious preventive action examples relate to machinery or equipment. Everyone can understand why regular oil changes will keep our cars running smoothly for a long time. It is harder to understand that, as a manager, you should give frequent and normal consideration to employees. They need maintenance also—not just to keep them from doing something wrong, but to reinforce the positive aspects of their performance. I know that sounds like the old "Theory X and Theory Y" argument (which is X and which is Y anyhow?), but I prefer to think of it as Baloo the Bear's perspective in the Jungle Book. Be positive.

Management style determines the ways that this positive approach will manifest itself. Find a sincere method of expressing interest in your employees and the work they perform without waiting for a catastrophe. Some managers are comfortable with scheduled monthly project reviews that are formalized and fully documented. Others spend part of each day out on the floor (management by walking around, anyone?). This allows them to interact informally and regularly with employees. They can develop a feel for the operation and the employees involved. These approaches, and any others that accentuate the positives, are not substitutes for normal performance and salary reviews. However, they can be used as a supplement that prevents or alleviates a terrible performance review. Think of it as insurance, paid in the coin of time and attention, to prevent a personnel "break down."

At this point, I can hear most of you thinking, "I am so

busy now with my job duties and responsibilities. How am I supposed to devote more time to giving my employees warm fuzzies and letting them talk to me about what they think?" As a manager, your decisions are only as good as the information you use to make them. Spreadsheets and trend analyses can only take you so far. Employees who feel free to voice their opinions, even the negative ones, will be much more inclined to warn you about what is happening on the floor. More often than not, where that first stitch should go is something someone other than yourself will know. Having that person want to tell you what should be done can be one of your strongest assets. To be sure, you will have to tolerate your share of "Chicken Littles," but look up anyway—the sky may actually be falling.

## Do Unto Others As You Would Have Others Do Unto You

As a consumer, I get rather upset that design engineers don't think about me when they create nifty new gadgets. Why is it that replacing the batteries in most portable electronic equipment requires fingers the size of toothpicks and the strength of titanium? Who decided that stacking AAA batteries three deep into a slot in a portable cassette player's underbelly would facilitate replacement? Did that person hope we would give up and buy a new unit? (I would like to thank the other designer who decided to add a little ribbon to pop the batteries out.)

The most baffling example is the procedure for replacing spark plugs in a Corvette. I cannot attest to all Corvette models, but one of my college roommates had to unbolt his engine mounts and jack up the engine to reach the rear spark plug on the passenger side. I came to understand that some Corvette owners hold this procedure in some reverence, as no one other than a dyed-in-the-wool 'Vette owner would appreciate having to do this regularly. In case you think I am picking on vintage 'Vette owners, I also know of a 94 model pickup truck that has to have the front axle dropped to change the starter. There is also the innovative "one price, no hassle car" line. It had to be redesigned after someone realized that the engine mount bolts could not be accessed after the body panels were put in place. I would have loved to see the mechanic's face the first time he had to ask a body and fender guy to replace engine mounts.

An Old-Fashioned Engineer thinks his or her designs through. He or she may even try to use and service the designs him/herself. This tends to eliminate the end user theatrics caused by designers who create equipment and processes they themselves have never used.

This ability is not limited to engineers. Many times managers come up through the ranks. They remember the more ridiculous company rules and regulations and work to change them. Managers who do not have this experience create a tremendous amount of inefficiency and work through ignorance.

When I was in college, I worked as a barback one summer. For those of you unfamiliar with the position, it means I was a gofer for a bartender. Collecting and washing glasses, cutting up fruit and preparing drink mixes were all relatively simple tasks. Changing out full kegs of beer for empty kegs in a narrow bar was a physical challenge (this was before $CO_2$ taps and central keg stations in a roomy, convenient location) but, for the most part the job was entertaining. I also learned a lot about people. I really enjoyed it until the fourth Sunday of my first month.

Every night, before I closed up, I had to clean the bar thoroughly. On a military installation, this means that everything was wiped down, including the gaskets on the coolers where the beer bottles were chilled. I was also supposed to fill the coolers with beer bottles to ensure cold beer for the next night's business. Unfortunately, my boss neglected to tell me that, on the fourth Sunday of the month, the inside of the coolers had to be wiped down with a bleach solution. This meant that, when I came in at 10:00 a.m. on Sunday morning, I had to remove all the beer bottles I had spent two hours stocking at 1:00 a.m. that same morning. I learned to break the rule of stocking the coolers nightly on the fourth Saturday of each month. That spared me a lot of work the next day.

## Walk A Mile in Another Man's Shoes

PERSPECTIVE IS WHAT YOU SEE COMPARED TO what others see in a given situation. Business is a chess game of moves and counter moves made to gain advantage over the competition. Many times managers have privileged information that guides their decisions. Unfortunately, their subordinates then see them as making decisions with little regard to the facts they know. If a manager is generally open with the employees under his or her direction, it may be adequate to state occasionally that privileged information is involved. Honesty, however painful or difficult to present, is still the best policy.

> Because engineers have specialized knowledge and vocabulary, they can easily fall into the trap of either talking over other employees' heads or talking down to them. This can be a problem.

Still, many managers find it easier to dictate direction to their employees (regardless of questions), rather than taking the time and effort to present plans and policy as clearly as possible. This is a two-way street. Employees then vie for opportunities to prove they are capable of making good decisions and understanding difficult situations.

Because engineers have specialized knowledge and vocabulary, they can easily fall into the trap of either talking over other employees' heads or talking down to them. This can be a problem whether it is inadvertent or a "payback" for a manager who is too domineering in his or her approach to project assignments. These games do absolutely nothing for productivity or achieving goals. They quickly become a detriment to an organization and should be avoided. Managers and employees alike should take time to listen to other's opinions. They should refuse to participate (inasmuch as their positions allow) in any games.

I don't believe there are many people who haven't heard the joke about the engineer yelling at the machinist in frustration, "What do you mean you made it to print?!" Engineers are notorious for getting caught up in tortuous details that blind them to obvious mistakes in their designs. Egos often precede common sense when systems and machines are being designed.

An engineer I worked with was asked to take a design for a tilting, glass installation table and modify it for smaller window frames. As I had created the original design, he felt the need to modify the table significantly, thereby putting his "stamp" on it. This particular engineer, being young and trying to make his mark on the world, was quite arrogant when dealing with the maintenance technicians and machinists charged with making his design's reality.

A couple of weeks after I had given the engineer copies of my original design, a maintenance tech stopped me on my way through the fab shop. He took me over to the machinist's bench to show me the drawings of the new table design. After a few minutes of intense scrutiny, I admitted that I didn't see a problem offhand. They laughed and pointed out several inconsistencies in part lengths and clearances that, upon their attempts to assemble the table, had become apparent. The movable arms that were supposed to hold the glass frame in place were so short that they could not be mounted on the base framework.

I knew the design could be fixed easily, so I asked if they had talked to the engineer about modifications. The laughter stopped immediately. In silence, they guided me to the back storage room for dead projects. On the floor sat the new glazing table, complete with short adjustable arms. I was quietly (yet angrily) informed that the young engineer had insisted they build his table as specified. He said he didn't want to be bothered by incompetent help who couldn't read a print properly.

It took many weeks and several hundreds of dollars to get a well-designed glazing table to the operators on the manufacturing floor. At that point, I decided I would listen to the fabrication people when they called from the shop with questions. Since then, I have had almost no problems implementing designs (even some with glaring errors) because I have let machinists and welders voice concerns and offer suggestions.

The only downside to this approach is graciously accepting some good-natured ribbing about my mistakes, paying bets I have lost when sticking to my opinion (usually a soft drink presented on the floor in front of witnesses), and having to reciprocate by presenting a fresh, engineering viewpoint and assistance when the fab people have ticklish problems. Not much of a downside, really—and the returns are immeasurable.

## If it Ain't Broke, Don't Fix It

I WAS CALLED ON THE CARPET for making this statement at a seminar. Apparently there is a whole new theory "out there," complete with books, authors, and gurus, that insists that you can fix something even if it isn't broken. Right.

My logic may not be perfect, but I fail to understand why, if a product, process or service is meeting the needs it was designed for, it should be fixed. People running around, creating work to improve items that are already working are perfect examples of why I felt compelled to write this book. I challenge anyone in any industry at any level to show me why resources should be used to analyze and modify a process (product, service) that is meeting its designed purpose, especially when, if I spent an hour in the same facility, I could identify and prioritize several processes that are not.

Remember the stitch in time that needs to be identified? There is a huge gap between allocating resources for a process that is *apparently* working right and allocating them for one that *is* working right. The old line about being "up to your butt in alligators" is more often true than not. I recommend focusing on those items within your control that will make you and your department truly efficient.

As an outside consultant, I am often placed in the awkward position of telling clients that there is nothing wrong with an area they want improved, and that their money and my time would be much better spent addressing areas that have serious shortcomings. Try explaining that, since Area X, which feeds parts to Area Y, is incapable of meeting Area Y's quality and quantity requirements, improvements in Area Y (such as JIT manufacturing, zero defects manufacturing, and continuous process improvement) will have little or no effect. In fact, Area Y will still be shut down 90 percent of the time.

I actually have seen companies chasing new manufacturing technology while their production lines sat idle because workers didn't have the proper wrench or screwdriver to adjust a machine. The problem is very basic: stockholders, clients, and boards of directors want their company to be on the cutting edge and in the forefront where the action is. The uninformed and often the unteachable entities in charge of the purse strings will approve dollars for a high profile project that smacks of the latest guru's preachings, but will balk at spending good money on something as mundane as screwdrivers.

An OFE&M manager or engineer's true challenge is to let glory-bound and high-profile, unbroken projects lie, while mustering the courage to tackle those dirty, slimy, and fetid projects that nobody wants to touch. Don't ever be lulled into thinking that if a project hasn't been completed for a long time, it must not be important. Many projects have a high level of difficulty but a low level of glory. That's why they're being ignored.

Learning to identify and prioritize projects is an art and a science unto itself. Try to remain objective and place yourself in the shoes of those who have to deal with the process daily. The Pareto rule is often condensed as the "significant few and the insignificant many" or the 80/20 rule. Most of your problems can be traced to a few select areas that need improvement. Don't waste time and resources on trivial pursuits, but jump right in and work on the true problem areas, even if it means getting your hands dirty.

*Many projects have a high level of difficulty but a low level of glory.*

I worked in a facility that manufactured architectural aluminum products. We operated a small 7-1/2-inch extrusion press and produced most materials in-house. The press operators were quick to tell me that the press couldn't be considered commercial because our production wasn't that efficient, mainly because our stretcher did not operate properly. The stretcher was vital; it ensured consistent dimensions and trueness in hot extrusions. Being the overly ambitious young engineer that I was, I promptly assigned myself the task of getting the stretcher up and running properly.

A brief investigation revealed that several engineers and the maintenance supervisor had looked into the same project over the years. It had proven to be an involved project that was invariably set aside after a week or two of scrutiny because only the press operators complained. When my boss found out what I was up to, he immediately told me to stop wasting my time. Upper management wasn't complaining, and there were other, more high-profile projects to be completed. At the time, I was the project engineer in charge of new product introductions for the plant (which handled the bulk of new product introductions for the corporation) so I was well aware of the predicament.

> An OFE&M manager or engineer's true challenge is to let glory-bound and high-profile, unbroken projects lie, while mustering the courage to tackle those dirty, slimy, and fetid projects that nobody wants to touch.

The extrusion press directly determined how effectively the rest of the departments operated. Therefore, I still believe that my prototype extrusions would not have been pushed back on the press schedule as often if our yield had been higher (which meant a properly working stretcher). Looking back, I now realize that overall plant productivity would probably have increased by as much as 10 to 12 percent had the stretcher been made a priority.

## Just the Facts, Ma'am

JOE FRIDAY, YOU ARE MY HERO. How many times have I reacted to bad news only to realize later that the news was in error? I lost count somewhere around 253. At what I assume to be about 637 (I'm a slow learner), I began to see a trend in my apologies for being overly excited about projects that had "blown up." After I gathered all the pertinent data (facts) and looked at them objectively, I invariably discovered it was not as bad as it first seemed. My blood pressure came under control when, instead of jumping up and down with the messenger, I tried to calm him or her down. This handy technique works equally well when you get good news. (Don't worry, it generally isn't as wonderful as the report first reads; again, deal in just the facts.)

This technique even works when the news is worse than it first appeared, even if you originally thought it was horrible. A calm, rational approach makes the gears of a solution engage much more smoothly no matter how fast the disaster train is bearing down on the spot on the tracks of dilemma where your 1952 Chevy pickup truck is stalled.

When I was a middle manager in charge of engineering and maintenance, I was called into the vice president's office to discuss water filter field failures. The matter was exacerbated by the fact that I had helped redesign the product, had designed the manufacturing work cell that produced it, and was the production supervisor in charge of the area. (I couldn't pass the buck on this one!)

> A calm, rational approach makes the gears of a solution engage much more smoothly no matter how fast the disaster train is bearing down.

Apparently, a bad product run had been shipped to Australia, where sales were brisk and field failures could not be tolerated. My initial reaction was "I'll fax our Australian sales rep a list of questions immediately. As soon as he replies, I'll investigate the possible causes of failure." That response gained me nothing but a tirade against my lineage and the threat to report my ineptitude and incredibly low IQ to corporate. I was told to find something I was capable of pursuing successfully (toilet cleaning was suggested) and that the VP would handle the situation himself.

I went ahead and faxed the questionnaire. Meanwhile, the VP made internal part design changes that required tooling changes and assembly changes and so on and so on. Within the week, 500 "good" water filters were shipped to Australia on a boat slated to arrive just in time to save the day. In the midst of this, my fax was answered. It seems that Australia can run as much as 120 pounds of water pressure in some residential areas. Our water filters were designed to operate at about half that pressure. The straightforward, low-cost fix was to provide inline pressure reducers in high water pressure areas.

Back in the plant, it was discovered that the design/tooling/assembly changes were now, in fact, producing defective products. I was then charged (in the privacy of my office this time) with changing everything back to the way it was originally. The 500 "good" filters on the slow boat were recalled, and another 500 new filters were air-freighted to Australia in time to truly save the day. I later estimated that this little exercise in not waiting for the facts cost the company about $54,000.

## Tunnel Vision:
## Can't See the Forest For the Trees
### (Or, A Paradigm Shift By Any Other Name)

PARADIGM SHIFT IS THE COOLEST CATCH PHRASE to come along in quite some time. I really like the sound of it a lot. Apparently, so do many others. Unfortunately, most people don't have a clue as to what it means. Tunnel vision and not being able to see the forest for the trees (both well-worn clichés) are much clearer but they don't sound as slick, so consultants can't charge as much to utter them.

Everyone experiences tunnel vision in some aspect of

their life. At its best, it can be very positive. If we get in a groove during the local bowling tournament and throw a perfect 300 game, we call it "focus," and we are held in awe and admiration. If, however, the next night we experience "gutter ball syndrome," and the groove becomes a rut, we are pitied and held at arm's length just in case it's contagious.

Managers, engineers, and production facilities (as entities) find a comfort zone, create a groove, and stay with what works in what becomes an "autopilot" mode. Changes in business environments, personnel, or regulations can rapidly indicate just how deep the rut has become. Sometimes it is so deep that the changes going on in the immediate vicinity go unnoticed.

Positive change can often be made quite easily and with little cost. Instead of paying big bucks to determine whether you are trapped in a paradigm, shift your personal or work habits slightly. Instead of arriving at the office each day at 9:00 a.m., come in at 6:00 a.m. for a few days. (For you early birds, try the inverse). Spend more time on the shop floor, in the office, with clients or in the marketplace—whatever disrupts your normal patterns. Change the clothes you wear. If corporate culture demands a suit and tie, wear a pastel shirt instead of one of your normal white shirts.

These changes in patterns will seem awkward unless you are one of those rare individuals who enjoy change. If you receive an inordinate amount of comments from your co-workers and employees because you have made these subtle changes, then it is time to analyze procedures and operations. If you are in a rut, it is more than likely that any department or personnel under your influence are also. In today's business environment, you must be making improvements constantly or you will lose ground rapidly. My favorite quote comes from a friend, a hot shot electrical contractor: "If you're waiting on me, you're backing up!"

In my experience, batch processes are most prone to paradigm. I cannot recall how many times I have been told, "It can't be done as a continuous process because of ...." The simplest example of this involves a tweeter (hi-frequency) driver assembly process. The lightweight diaphragms were glued and put on cookie sheets in batches to dry. A cart then moved them around the bench to the other side for the next operation. Two operators were positioned so they could talk to each other across the bench. The second operator often was left waiting while the first operator filled a tray. Naturally, productivity suffered and quality problems weren't discovered until at least twenty-four units had been assembled and tested. Conveyor systems had been deemed too bulky to fit in the area or too rough to handle the delicate voice coil/diaphragm assemblies. Cost, as usual, was also a major consideration.

The solution proved to be very simple. I purchased a length of smooth downspout at the hardware store and cut a short length of it to cross-feed the diaphragms from one operator to the other. The major difficulty was determining the angle at which the parts would slide all the way through, yet not pick up enough speed to damage each other at the discharge end. It turned out that the assemblies didn't need to dry for more than a second or two. I combined several of these simple "paradigm shifts" and, for about $500 and two weeks of implementation time, yielded a 45 percent reduction in rejects while going from 125 to 525 units per day. (Okay, I couldn't recall these numbers, so I looked them up in my old notes. See how helpful documenting everything can be?).

Remember, if you can't see the forest for the trees that doesn't mean you need to cut down all the trees. It just means you should consider changing your perspective or focus.

## Eating the Elephant One Bite at a Time

I'D LIKE TO MEET WHOEVER UTTERED THIS phrase for the first time. It has a tremendous visual impact even on non-vegetarians.

Most important changes in our lives as managers or engineers come with a heavy burden of data that must be assimilated in a short period. Vernacular associated with a new process or product can be daunting enough, never mind learning the associated concepts behind the new words. Compounding the situation is the perception (right or wrong) that managers and engineers have an innate ability to operate on a higher plane and instantaneously understand complex new situations.

At this point I will assume you are not one of those individuals that promulgate this myth by studying behind closed doors on evenings and weekends just to appear smarter than the average bear. If you find your personal comfort level being trod upon by the changes swirling around you, then take a moment and focus on the word or phrase that seems the most confusing. (Starting at the most difficult point means you're headed downhill on the information highway for the rest of the trip.) Then pick that word or phrase apart and put it back together again.

When I was an undergraduate student, I was that guy in the back of the lecture hall who constantly interrupted the professor and asked him to define every other word so that I could make sense of his sentences. I reasoned that my confusion started with words, led to sentences, and then to complete paragraphs until I could not understand the entire lecture. Instead of sitting in agitated silence, I agitated the professor instead until he used more familiar terms. Most of my professors, as my grade point indicated, thought I was an idiot. My fellow classmates, on the other hand, thanked me after class. They encouraged me to continue making them look good by asking the "stupid" questions they were too embarrassed to voice.

We managers need to do two important things: (1) ask questions when we don't understand a situation, even if it means opening ourselves up for ridicule; and (2) encourage our co-workers and employees to ask questions. Once we understand a situation, we must always be open to signs that others around us are still unenlightened. At that point, we must serve up the elephant in palatable, bite-size pieces that allow for positive forward progress. Patience is also a great quality to call upon when you are tackling an elephant-sized problem. Let's face it, an elephant is at least a twelve-course meal.

> If you find your personal comfort level being trod upon by the changes swirling around you, then take a moment and focus on the word or phrase that seems the most confusing.

My serious avocation is writing and recording music. Basic recording with a multi-track cassette machine is relatively straightforward. When I first started out years ago, I added to my musical toy collection very slowly—usually one major purchase each year. Since I am both an engineer and a techno-nerd, I was embarrassed when I realized that my latest wonderful equipment purchase was not improving my recordings, but was instead intimidating me to the point of inaction. By the time I had the funds for the next new piece, enough time had passed that I was operating the old equipment almost by instinct. When I was offered the better part of a professional digital studio at an unbelievably low price, I gleefully made the quantum leap from my amateur set-up. What I failed to take into consideration was that the techno-shock caused by one piece of new equipment would be amplified many times over with this giganto purchase.

I unboxed the equipment and stared at it in stark terror for several days. The manuals did nothing to ease my concerns, because the switch from analog to digital recording involved a host of new techno-terms and concepts. These threatened to overwhelm my ego and put an end to my recording sessions. As a last resort, I decided to start with something I knew how to do. I plugged all the equipment in. I pushed the buttons marked "power" and watched the LEDs light up. Then I got bold: I connected some of my old, familiar equipment to the new equipment with old cables and cords I knew wouldn't hurt me. Today, I can place you squarely in the middle of my studio control room (with enough buttons, lights, and switches to make a NASA control engineer green with envy) and calmly explain what everything does. Yes, I am recording again, and it is much better than before.

The point is this: my music equipment is no more or no less intimidating than the automated equipment or control procedures I encounter daily in my consulting business. It actually shares much of the same technology. What you must remember (and this is hard) is that, once you eat your personal elephant and help your employees ingest theirs, anyone new to the situation will face an elephant easily as large. In fact, it may be the exact same elephant! Be careful how it is served up.

## The Early Bird Gets the Worm

MY SPIN ON THIS WELL-WORN CLICHÉ may disappoint diehard, up-at-5:00 a.m.-types who disdain associating with those of us who are generally dysfunctional before noon. Getting the worm by being the early bird has very little to do with the clock. It has much more to do with your attitude about jumping right in and tackling difficult jobs.

In my experience, I often hear people say, "I wish I had called earlier this week about…." You yourself have probably run into individuals or corporations who take months or years to implement a new product or process simply because the person in charge is waiting for the proper moment to leap into the unknown. If you haven't, look up your local chapter of inventors and hang out with them for an evening. You are guaranteed to hear at least one story of how Jim designed the NASA lunar rover but wasn't sure mankind would make it to the moon, so he didn't build the prototype or file a patent on it. The scary part of these stories is that most of these people can pull out paper napkins dated May 5, 1954, that feature full scale dimensional drawings with specifications!

Too many times success hinges on the ability to act without fear of failure. Just Try It.

I have butterflies and a touch of fear every time I start a new engineering design project. Am I really creative enough or talented enough to accomplish this? After many years completing difficult projects successfully, I started asking other designers if they had the same fear of failure. These were people I hold in high esteem because they create and mold ingenious solutions. Without exception they laughed, looked away, and then stared straight into my eyes. Most replied that they even go beyond butterflies and to the brink of doubting their own sanity.

Too many times success hinges on the ability to act without fear of failure. I believe that, if I keep enough irons in the fire, sooner or later one of them is going to get hot enough for branding. Success comes to those individuals or corporations who jump in feet first. I'm not advocating lack of preparation (see "Do It Right the First Time"), but hiding behind the veil of caution only creates an excellent vantage point for watching the world go by. Nike has the phrase "Just Do It." I prefer "Just Try It."

Being the first to jump in when difficulties present themselves will give you confidence even when you fail. I know an engineer who was part of a million-dollar robotics project failure. The main lesson he learned was that he didn't die and the world did not come crashing down on him because he failed. Now he is so successful at implementing projects and so arrogant that it is difficult to be around him long.

Managers are so often rated on their ability to implement projects successfully that they create an environment that does not tolerate failure. As a consequence, many employees don't do anything that could lead to failure—let alone success—because they don't want to risk losing their 3 percent annual raise or possibly even their jobs.

> The only thing harder than motivating yourself to try the unknown and risk failure is inspiring others to do the same.

When I was a junior project engineer, I loved getting things done and seeing my ideas spring to life from the drawing board. (This was, and still is to a large degree, a very politically incorrect approach to corporate life.) One day, a senior corporate Vice President visited the plant to observe firsthand the progress we were making on a new product introduction. He conversationally asked me how things were going. I replied, "I guess they're going okay, because I haven't made anyone angry today and I'm not currently in trouble."

His response startled me and gave me insight into how the corporation was operated. "Well, don't worry about being in that kind of trouble; it just means you are actually accomplishing something and getting it done."

He knew I was the engineer responsible for the product coming on-line ahead of schedule. He also was aware, as I was not, of the managers over me struggling to take credit for this. The reason I got in so much trouble—20/20 hindsight is wonderful—is that I could answer corporate vice presidents' questions extemporaneously while the middle managers who claimed credit hadn't done enough work on the project to know anything about it. The only reason I got most of the projects I did was because I was the only person willing to risk failure and tackle them before they became a sure thing.

The only thing harder than motivating yourself to try the unknown and risk failure is inspiring others to do the same. Management by the "go-for-it-I'll-support-you-even-if-you-fail method" is not as easy as it might sound. We are so well programmed to avoid failure that it is almost impossible to convince employees that yes, even if they fail, they will still be supported. Only a history of doing so will give your employees any faith in you and your word. The first few projects they undertake with a high expectation of failure will be like pulling teeth without Novocain. Don't expect any volunteers.

The most interesting and probably the best example of this management style came from a Karate instructor I had as a teenager. Sensei Jones taught Karate for kids at a military base youth center. Most "military brats" live with a very high level of discipline. Although most rebel at some point, when it comes down to brass tacks, we are used to doing as we are told by anyone in authority. Our instructor took unique advantage of this. He had one basic rule in the dojo, or school: we had to try anything he asked and we could not say, "I can't." He never ridiculed us for failure nor made examples of our mistakes. On the contrary, he would only yell or discipline us if we did not try a new technique or uttered the dreaded, taboo phrase, "I can't."

The first few months he taught were a nightmare. We were constantly pushed to try new and more difficult techniques that intimidated us. After a while, we noticed that we were never asked to deny our fears or concerns. We also were never asked to try anything our instructor didn't feel we were ready for. Once we developed confidence in our instructor, we began to volunteer to assist with new techniques. Our learning curve consequently was abbreviated and we progressed rapidly in our lessons.

Not being paralyzed into inaction by uncertainty is a desirable attitude to cultivate in yourself and employees. A manager must be aware of the skills, education, and experience of each employee so that an unnecessary burden is not placed on those who trust the manager. The added responsibility the manager must bear ultimately reaps big dividends and justifies the effort in the long run. The lessons I learned in that Karate class all those years ago still define my success.

## Give A Man A Fish

GIVE A MAN A FISH and you feed him for a day. Teach him to fish and you feed him for a lifetime. No problem, unless his success at catching fish determines whether or not you get to eat. It is difficult to teach someone the finer points of fishing or laying out a manufacturing facility and continue to be productive at the same time. There is the real possibility that you will tangle your fisherman in the net and toss him overboard in the next cast. It is nerve-wracking, difficult, and one of those wonderful things that usually falls under the "delayed gratification" heading.

While setting up an automatic welder for a new product line as a young engineer, I was assigned a quiet and very intelligent line worker named Bob. Bob readily admitted that he was not sure how to program the PLC (Programmable Linear Controller) that was the heart of the system. Because I was so enthusiastic about the new equipment, I inadvertently did the right thing. As soon as prototype parts became available, I taught Bob the basics of programming. Having someone look over your shoulder never helps when you're learning something new. So, I indicated the stack of expendable parts and explained the different weld configurations. I then left Bob to his devices while I tried to get a handle on the other work cell aspects. Several days later, when I had some "free" time, Bob told me he was out of test parts. I took a look at the stack of parts and was amazed. Naturally, there were many pitted, burned, and misaligned parts in the stack, but it seems that Bob got a little carried away while learning the new gear. Since he didn't know the design limits, he proceeded to produce higher quality and tighter toleranced welds than the engineering guys (including me) had assumed were possible. This stood us all in good stead a few weeks later when the salesman in charge of the product introduction decided that, for aesthetic reasons (my personal favorite), the welds needed to be prettier and conform to tighter tolerances. I've never seen a more consistently pretty structural weld than those Bob could produce with that custom welding set-up.

Your success as a manager will depend greatly on your ability to teach your employees how to be successful in their own endeavors. I had a mentor who warned me very early in my career of the pitfalls of being a manager with my personality traits. I am usually detail-oriented, creative, and meticulous when executing projects. Unfortunately, such a manager may not have employees that approach problem-solving the same, with the same execution speed, or the same degree of finesse. Those of you who are fast and loose problem solvers, face the same problem, but on the opposite end of the spectrum. You will be equally frustrated with a slow moving, detail-oriented employee. Managers should define problems and what constitutes a successfully completed project and then step back, giving direction only as needed and with few editorial comments.

This does not mean you throw your employees to the sharks in chum-infested waters by walking away and returning just before the deadline to see how the project came out. It means keeping your finger on the pulse of the project and gently nudging it when it appears to be heading irreparably off course. Being aware of your personality traits and those of your employees is critical. Knowing your employees' (and your own) strengths and weaknesses will allow you to give them a starting point that in turn creates a solid foundation for whatever problem-solving approach they use. For those employees who are more literal in following instructions, be sure to define the project steps clearly in terms they know. For those employees who interpret your directions more creatively, give them more latitude for stimulating new avenues (some of which you may never have dreamed of). In both cases, be very clear about what constitutes success.

> Your success as a manager will depend greatly on your ability to teach your employees how to be successful in their own endeavors.

My greatest success as a manager came to me out of the blue and thus was a total surprise. I planted the seeds of opportunity in a handful of competent individuals, and they went far beyond the boundaries of success. When I was a middle manager at a medium sized manufacturing facility, I noticed a gap in our management structure. I theorized that using "lead" people as entry level managers could close that gap. Already there were individuals voluntarily doing the work because they naturally tended to do so, but they had no training or support. I conducted informal interviews to determine how well a lead training program would be received. The initial negative response surprised me. Several of the other clichés I've already discussed explained the reticence (do unto others as you would have them do unto you; walk a mile in another man's shoes; tunnel vision; can't see the forest for the trees; and eating the elephant one bite at a time.)

Determined not to be discouraged, I presented them with a confidential and hypothetical "what if" scenario and asked them to write an informal essay for my eyes only. (Fortunately, I was respected as a person and a manager, so I got a much better response to this query than if another manager had made the same request.) I asked each individual to write his or perceptions of what was wrong with the company, the products, and management. They also had to include solutions for each problem they identified and define the most effective role they could assume in addressing these problems.

Talk about impassioned essays! I don't think anyone had ever honestly asked these key people for their opinions. In addition to giving me specific ideas of the areas that needed improvement, they also provided me with a blueprint for the support training. I had very willing students in informal classes on keeping records and interpreting data related to production, maintenance, quality control, people skills, and project cost justification.

I enjoyed working with these people, but my own personality was working against my position as a middle manager. I was butting heads with upper management and trying to be a buffer for the people who worked under me so that they would have the time and space for achieving their own personal goals. Not surprisingly, I was asked to leave the company after a few short years, and I started my own engineering firm at that time. (It became much harder at that point to argue with upper management, although I have fired myself several times since then). That was that, as they say, and I left the past to lie and moved on.

Several years later I was training a young engineer in my company to do plant layouts, and I noticed that we were only a few blocks away from my old plant. Not having burned any bridges (at least not beyond repair), I stopped in and asked for permission to show him the facility, as it would be good experience. Little did I know how good an experience it would be and for whom.

I was amazed from the moment we entered the production space. Physical changes in the plant, including housekeeping, lighting, safety, and product flow (all of which I had fought for and mostly lost), were readily apparent. Almost without exception, the lead program people had been promoted to positions with more responsibility, including supervision. I received a mild rebuke from one former employee (water filter production lead) for leaving the company, because all lead training had at that point stopped. She was very pleased to report, however, that she had been promoted and was in charge of quality assurance for the plant.

My biggest and most unexpected compliment came from my former maintenance lead, now maintenance supervisor. While examining his new and meticulously maintained equipment, I asked him how had he accomplished so much in such a short period of time. Because he was a very private and quiet person, his response almost floored me: "Someone once taught me that documentation and justification are the keys to getting what you want." His smile and handshake at that moment still mean more to me than all the money I made at that company.

I was blessed with the opportunity to view the results of my managerial efforts in a three-year time warp. The success of the people I helped train was sweet indeed. They had learned not only how to fish, but how to build and repair nets to feed those around them.

## Pareto and His Rule

PARETO'S LAW IS ONE OF THOSE well-known axioms that everyone can paraphrase and, for the most part, understands in theory. Then reality rears its ugly head, and theory goes out the window. Sometimes referred to as the 80/20 rule, Pareto's law refers to the tendency of most systems to be defined by the significant few and the insignificant many. (I have it on good authority that Pareto actually had little to do with this theory.) It is safe to say that this rule applies to an automobile when you consider how many parts are critical to a car's movement (relatively few), versus how many parts are required to make the car more efficient or comfortable (except for dune buggies which don't apply in this example). In most any system or problem encountered, this holds true.

The most obvious example of Pareto's law in action, from a management view, is a plant start-up situation. For those of you who have not experienced a plant or product start up from the ground up, it generally follows the following simplified outline:

1. Hire a project manager to oversee all operations.
2. Hire a small staff that will wear lots of hats.
3. Build the shell of the building or product line.
4. Buy only enough hand tools necessary to do the basic assembly processes. Bootleg assembly jigs.
5. Hire and train a bare minimum of line workers to produce prototypes and initial production.

*Production begins in earnest, the learning curve is ramping up, and productivity is weak.*

6. Hire more line workers to get production to required levels. The learning curve is flattening.
7. Build automation and detailed jigs and fixtures to raise productivity.

8. Hire more managers to help ease the stress caused by overseeing lots of equipment and line workers.
*Product introduction surge sales flatten as production learning curve levels out. Productivity is high.*

Then, just when the automation is debugged and fully on line, and the line workers are running in full swing with individual department managers overseeing multiple line supervisors, sales level out and drop. Now the Pareto rule comes into effect. In step 5, the plant was running well, but productivity was not at its peak because the learning curve had not yet peaked. It would seem quite obvious at this point to downsize from position 8 to position 5 now that sales have flattened, and productivity is at its peak.

In truth, only very desperate or very aggressive companies are doing this. Many factors affect the decisions. If people are laid off because of new automation or production increases, even if they were hired as temporary employees, a negative image of the corporate identity is the result. It worsens if reducing the number of line workers eliminates permanent staff and managerial positions. Unions add a completely unique twist to the equation that I don't even attempt to understand, let alone embrace. Arbitration anyone?

I have found that visualizing a company as a start-up, undercapitalized, and struggling-to-grow entity is the easiest way to identify the necessary assets. If you are running a mature company in trouble, then treat it like a brand new company and apportion assets accordingly.

The trick is determining the critical people/parts and putting resources in the areas where they can do the most good. PERT (project evaluation review technique) was developed to address this situation: it identified items that must be addressed for a project to be completed successfully. This is called the critical path. People and the positions they hold within a corporation are (or can be) much more difficult to assess than nodes on a chart, however. Many politically adroit individuals can justify six weeks of vacation a year, a six-figure salary, and all the power lunches they can schedule, while not actually making any positive contributions to the corporation. Find a way to get these people into marketing or the unemployment line, but keep them out of research and development, and don't let them go into politics.

The long-term result frequently seen in the United States is a bloated and overweight corporate structure struggling to remain upright on spindly legs. Many times the structures were laid out during either start-ups or growth-and-glory periods when floods of money were handled with buckets. When the money becomes a trickle, and downsizing is necessary, more often than not, the vital organs are cut out and thrown away instead of trimming the fat. Politics and a true inability to determine the significant few are often the reasons.

> Old-Fashioned Manager treats every decision as if it affects his or her personal life.

I have found that visualizing a company as a start-up, undercapitalized, and struggling-to-grow entity is the easiest way to identify the necessary assets. Get back to step 5 on the list, or better yet, don't progress past that point in the first place. Wait for the learning curve to flatten out completely, and let the new introduction sales figures level out. If you are running a mature company in trouble, then treat it like a brand new company and apportion assets accordingly. Remember, an Old-Fashioned Manager treats every decision as if it affects his or her personal life. Compassion and understanding can make implementing the hard decisions easier, as long as you are sure your analysis is correct.

Pareto's law applies just as well to traditional engineering problems. In a cold-forging operation, many factors can affect the forging quality. If insufficient fill is a problem, then blank size, blank configuration, press tonnage, or die clearance are significant factors to consider. Although many things can contribute to poor fill investigating the quality of the micro finish on the blanking dies or the blanks' temperature variation before considering blank configuration doesn't make sense.

Often, it is easier to investigate blank temperature variations (especially if the department just purchased a brand new digital thermometer that is accurate to within .00001° Kelvin, than it is to investigate die clearance (especially if it means confronting a belligerent die setter). Know your processes or talk to someone who does and go with the odds. Occasionally, a normally insignificant item will cause significant problems, but the instances are rare and they make great fodder for coffee break discussion.

# CHAPTER 7

# Common Traps and Pitfalls

As a practitioner of OFE&M,
don't fool yourself into believing that you
can make fundamental changes in anything
other than the approach you personally take
towards your work and the
environmental factors you encounter
throughout your day.

As a consultant, I am in a unique position. I see a wide variety of combinations for the *Rule of Threes Matrix*. Individuals, departments, and corporate identities come in all shapes and sizes. Sometimes the permutations are successful, sometimes not. I have even seen the matrix yield hilarious combinations of working personalities that were also very successful. Unfortunately, this is rare, and almost everyone I know can much more easily recall or relate those RTM combinations that were disastrous or, at best, dysfunctional.

Many prevalent attitudes, myths, and outright lies exist in the matrix components and cause problems. These vary from individual personality traits to generally held societal beliefs that hamstring success. As a practitioner of OFE&M, don't fool yourself into believing that you can make fundamental changes in anything other than the approach you personally take towards your work and the environmental factors you encounter throughout your day. Remaining aware of traps and pitfalls will allow you to chart a different and safer course.

Who knows where to start? I guess one of my favorites is the "who wrote the book" syndrome. I love the irony here. Because you are reading this, I have now "written the book," and thus am now automatically elevated to expert status.

While working on a joint project with a federally funded engineering team, I met a Ph.D. candidate who had written the book on quick-change die tooling. We were having a planning meeting with him before his first site visit to the client, and I wanted to know where he got his experience. As an ice breaker, I asked him if he remembered to bring his steel toes for the plant visit. (For those of you who have not worked in heavy manufacturing facilities, this means steel-toed shoes, a required personal protective safety equipment item.) His puzzled look led to one of my own when he asked me what I was talking about. He didn't know the slang term for steel-toed shoes! Furthermore, he didn't own a pair and didn't intend to waste his money on such a ridiculous item. I was flabbergasted. This author claimed to be an expert in quick-change tooling design, yet he had never been in a plant or machine shop that used or made tooling. His expertise came from reading books about the topic.

While I love to read and have a great reference library, I do not advocate this kind of expertise. I don't care how many books he's written, I still wouldn't take advice from a tooling expert who had never had greasy hands. It would be the same as buying software written by a computer expert who had never actually used one. Theory only goes so far.

Being in business for myself, I sometimes want to laugh and cry alternately at the smoke and mirrors consulting firms employ. A true consultant has expertise and provides solutions, not general guidelines. I had a high school history teacher who marked our term papers with the initials "gg" in bright red ink. This stood for "glittering generalities," and it was the kiss of death if you received more than one on a given paper. This was her way of letting you know that she knew you hadn't researched the topic and had added copious amounts of filler to reach the number of words required.

I don't care how many books he's written, I still wouldn't take advice from a tooling expert who had never had greasy hands. Theory only goes so far.

Most big consulting firms operate just this way. They offer really nice report covers, incredible color graphs, and multitudinous charts to outline and diagram the way to success. No actual content, but lots of glitter.

Some of you may heartily disagree with this observation about big consulting firms, but my research gets even better. One of my junior engineers was visiting with a college buddy who went to work for one of the Big Six consulting firms. Ralph (again, the name has been changed to protect the competent) was relating stories of design projects he had implemented while under my tutelage. Included were stories of what a pain in the posterior I could be and how much he was learning in such a short time. His masters' degree in engineering hadn't really prepared him for all aspects of onsite project implementation, but he found the work challenging and rewarding. He reveled in seeing projects from the drawing board come to fruition. He was proud and rightfully so when the client heaped praise on him for taking a difficult problem and addressing it beyond expectations. His friend, Bob (I've even changed the names of the incompetent to avoid embarrassment) was quick to point out that his experience was quite different. His big consulting firm operated at a much "higher" level, doing strategic planning and guidance without actually bothering with any project implementation. He finally put it in these terms: "We don't do any actual work. We just tell the client what needs to be done and let them find somebody who can do it."

I've heard the complaint about consultants announcing there's a problem (which the client already knows), but not giving any concrete solutions (which the client needs and does not have) so many times that I think it should become a cliché. I do, however, appreciate these kinds of consulting firms; I get a great deal of my own work this way. It means I have to leave my Gucci suits at home and put on my blue jeans instead. I like it that way, because I get things done. If you are an engineer or manager working for a corporation, try not to fall into the glittering generalities trap. It is possible to plan strategically and implement.

If you find yourself trapped in the ivory tower of design, get your hands dirty and hike down the paper trails you create to see where they lead. (This is doubly true if you actually design equipment, or if you are a manager that rarely faces the product of your own policies.) If they go nowhere, you know you need to redesign them. And don't worry, those paper cuts will heal quite nicely over time.

I strongly believe that many of today's business and manufacturing problems originate at universities. The emphasis on grant dollars and the "publish or perish" mandate leaves precious little focus on learning. As a graduate teaching assistant, I was shocked at how many students asked me to explain material outside my area of expertise. It seems my interest in teaching outweighed my lack of knowledge in any given subject. Compare this to full professors with tremendous expertise, but no motivation to teach. Tenure, pay raises, and pecking order among faculty members have nothing to do with teaching ability. If an instructor's prevailing attitude is "Why should I care about teaching?" it will be mirrored by the student's "Why should

I care about learning?" attitude.

I remember laughing at many of my classmates' gyrations as they chased test files rather than studying the raw material and figuring it out for themselves. On test day, if the answers weren't already memorized, then a great deal of note sharing occurred to ensure a high test score. I also remember many of my classmates wondering why I refused to cheat to maintain a high GPA, especially since I was on a partial scholarship that had minimum GPA requirements. The truth is, I was deeply concerned that I would be humiliated in the "real world" if I did not have the knowledge necessary for tackling the engineering projects that would be assigned to me. Little did I realize that the patterns of cheating in college (or grade school for that matter) continue past graduation. Apparently, it has gone out of style to work for a living. I don't just mean back-breaking manual labor, either. If a problem requires too much research or thinking it seems to be shelved in favor of easier projects. A lot more effort goes into creatively side-stepping issues than would be needed to address them directly.

The result is this: in any organization you can identify a handful of individuals who produce a significant amount of the department's or division's useful output. The rest of the individuals are still coasting, applying their efforts to getting a promotion or a corner office with windows. The "Big Bucks and Short Hours" myth promulgated by most universities is deeply imbedded in our work habits.

I have both taught at and been asked to speak at several universities. I realize that the prevalent attitude, "If I get a college degree, I will be worth more and not have to do manual labor for a living," is somewhat true, but I do not understand why it's carried to such extremes. I have listened to inexperienced college grads bitterly complain to professional societies' senior members about the mistreatment they receive at the hands of uncaring companies. Long hours (45-50 hours a week) and low pay ($30,000 annually) are often brought up, because tales of sugar plums were oft repeated to students struggling with tuition payments by counselors and instructors. I don't know very many green grads worth $45,000 a year for a 40-hour work week. To hear students tell it, though, they are all but promised that a choice of jobs at any pay rate, work level, and location in the world will be theirs for the asking once they obtain their degree.

These are the same students who are asked to regurgitate memorized material on request and have no clue (nor do they care) how much real money a company must generate to pay their salaries. Beware of those individuals who truly believe they deserve better simply because they want more. This attitude is not restricted to college graduates, not by a long shot. I have heard the line that, because of time-in-grade (the amount of time in a given position) an individual should receive more money as a matter of course, the same way that vacation weeks typically accrue. No mention is made of plans to acquire a diversity of skills or a depth of knowledge to benefit the company. The "gimme" generation has entered the work force, and it ranges in age from 16 to 60.

A college student made water filters for me during her Christmas break one year. She only knew from other department members that I was that particular assembly area's supervisor. When I checked on her training progress and her ability to work as a work cell team member, my area's lead and more experienced line personnel were full of smiles. It seems that this particular young lady was concerned that she was getting dirty. She also thought that the other members of the team worked too fast. She then confided in me, well within the hearing of the other cell members, I might add, that she was glad she was in college, would get a degree, and would not have to work this hard in conditions like this ever again. She asked me if I had ever considered going to college. This brought on a good-natured round of laughter from everyone, including me. The other members of the cell enjoyed telling her that I not only had a college degree, but was the manager in charge of engineering and maintenance. Her mortified comment was, "You mean you actually went to college for this?!" She lasted three days and was never heard from again. I hope she is sitting at a desk now and doing a job that meets her expectations.

To prevent total despondency over today's workforce, I would like to inject a positive note here. I have spoken at a few universities that have actually asked me to return and give students and even professors a reality check. Several students have asked advice on dealing with instructors who teach by rote. Some managers and line workers have even asked my advice on the different paths they can take to achieve success. A realistic work ethic is alive in small pockets. I sincerely hope that from these wellsprings of hard work will flow many a successful Old-Fashioned Engineer and Manager.

Another malady I see frequently is the "Tools vs. Toys " syndrome. We managers and engineers all use techniques we've been taught and picked up along the way. Sometimes we use the shiniest tool we have because we like it, and it worked well once. Often, the best tools in a mechanic's toolbox are the ones that are slightly rusted and bent, or even homemade. They obvi-

The perfect application for the perfect tool is rare, and force-fitting a solution to a problem is the quickest way to lose credibility.

ously are used often, even if they didn't come with silk slipcovers and special wooden cases. Glamorous and well-publicized theories are usually nothing more than repackagings of things we already know minus the glitz, smoke and mirrors.

Chasing after the perfect tools can cost a lot in terms of money, time, and sometimes reputation. The perfect application for the perfect tool is rare, and force-fitting a solution to a problem is the quickest way to lose credibility. Laptop computers and simulation modeling are a couple of tools that, when used properly, can save time and money. However, don't be fooled into thinking spending the time and money on acquiring these tools will instantly make you successful. I have often done a great deal of design work and "what if" exercises on a napkin over lunch or on an airplane. The cost was minimal (the flight attendant didn't ask me to turn off my pencil or my brain for fear of causing a crash, either) and the results were generally good. Sometimes a problem is complicated and needs the "big guns," but if it is not, be prepared to raise the ire of all involved if you waste energy and money playing with your toys while the guy at the next desk pulls out a napkin with the solution on it.

> Glamorous and well-publicized theories are usually nothing more than repackagings of things we already know minus the glitz, smoke and mirrors.

A perfect, if not embarrassing, example of this occurred when my college roommate and I were casually doing a few dozen thermodynamics problems one afternoon. Thermo is a subject that is more conceptually challenging than it is mathematically challenging—once the student understands how to approach the problem, solving the equation is pretty straightforward. Anyway, George (this name has definitely been changed as George is one of the most intelligent people I know) started cussing his calculator and slammed it on the bed. It bounced quite high, but landed unharmed. George was mad, but being really smart, he threw it at something soft to protect his investment and still vent his frustrations. I was about to murmur condolences about the tough homework when he looked at me incredulously and explained his actions. He had just realized that he had added 2 and 2 on his $200 calculator and it had given him the correct answer of 4. I started to laugh but then realized that I myself had just added 1 and 4 on my calcula-

> One of the most basic OFE&M principles is to use what you already know.

tor. I had also gotten the correct answer, but was equally mortified to realize what we were doing. To make amends, we solved the remaining homework without our wonderful toys. It took awhile to do the long division, but we felt better knowing that our calculators were being reserved for cube roots and permutations rather than addition and subtraction. Come to think of it, I believe George even put away his mechanical pencil and used a wooden, no. 2 pencil for the occasion.

The point is this: our calculators were wonderful tools, but much slower than other approaches we had, namely the dreaded math tables we all memorized as kids.

Remember, one of the most basic OFE&M principles is to use what you already know. In other words, don't discount your knowledge just because you already have it. You should always question why you feel the need to go to that seminar, take that additional college class, or get that higher college degree from "Big Name U." If you feel you need more serious knowledge, consider all your options. College classes, trade school classes, or a full-blown degree program may then be in order. However, if you want to pick up a new tip or two, a seminar taught by someone with real world experience may help. Avoid seminars taught by college professors who specialize in theory or experts who wrote a book, but have no hands-on experience. The major complaint I hear (and have experienced) about those seminars is the small return on what is usually a big investment because the individual attending the seminar already knew more than the instructor!

Chasing more knowledge or information to make a "truly informed" decision can be a procrastination smoke screen used to hide a fear of committing to a solution that may or may not work. One of the more subtle and devastating problems I have encountered is procrastination.

Many people function better under deadline pressure and accomplish a great deal that way. The individuals I am addressing, however, are those that make great plans and keep waiting for the right moment to implement them. Often, their procrastination can be traced back to a fear of failure (or even a fear of success) that paralyzes them.

If you set a goal and attempt to reach it, one of two things will happen: failure or success. Failure, in this context, simply means admitting a mistake, analyzing it, and correcting it. Success means having to reset goals and mile posts and then starting over again at a higher level. I sometimes think that success, at least for me personally, can be more difficult to handle than failure. I have learned how to evaluate problems and setbacks, and I am now to the point where I can take them in stride. A success, however, can be a relatively rare event that doesn't have a clearly defined end point. At what point should you stop reveling in the success and reevaluate the path? Will it take you higher (assuming you don't want to take time to coast and relax after the uphill climb) or do you need to choose a new path? The fear of the unknown can be the same with either success or failure. If

Failure simply means admitting a mistake, analyzing it, and correcting it.

you are an OFE&M practitioner, you will always be able to deal with both sides of the coin. Simply look at where you are, what tools you have available, and what tools you would like to acquire. The path will define itself.

The sooner you attempt a project, the sooner you can determine if you possess the tools to implement a solution successfully or not. I have met several intelligent individuals with more knowledge in a given field than I have, yet I implement projects more effectively. My style is to check water depth and temperature, look for snakes and alligators and then jump in! I am swimming around and generally enjoying myself, while they are still on shore determining the water's pH, total suspended solids, and bacterial content. I have been known to cut my foot on broken glass a time or two, but no venture worth undertaking is without risk.

Of course, there are other resources besides your own brain and classes. Coworkers, libraries, and the Internet can provide a quick answer to a properly posed query. Asking for help is not tantamount to screaming at the top of your lungs, "I am stupid and don't know anything!" Make sure your ego isn't controlling your actions to the point of ineffectiveness. "I am smart enough to know what I don't know" is a perfectly legitimate stance.

*The path someone else takes need not be the same as yours if he or she reaches the mountain top safely.*

Ego. I'm not sure anyone's ego is larger or more ungainly than mine, so I guess that makes me the expert category on handling the problems ego can throw into your path. (Somehow I think I just stroked my ego while at the same time admitting it is a problem!)

"Ego-added" versus "value-added" management will put you on the fast track to getting sidetracked. If you don't need to put your finger in the pie, don't. My first business mentor warned me that, as my career progressed, I would have to learn to understand that other people don't approach or solve problems with the same attention to detail as I do, nor will they use the same techniques. The

trick is to recognize whether goals are being met and concerns are being successfully addressed. The path someone
else takes need not be the same as yours if he or she reaches the mountain top safely.

Design engineers and project managers are susceptible to
thinking "that is not the way I designed this to work." I have
actually seen brilliant engineers
become angry to the point of
speechlessness about changes
being made to their computer-
simulated facility years after the
plant was up and running. Yes,
the original simulation provided
important and pertinent informa-
tion, but no one has a crystal ball so accurate that they can
design an entire facility and not expect any changes over time.

"Ego-added" versus "value-added"
management will put you on the fast
track to getting sidetracked. If you don't
need to put your finger in the pie, don't.

Ego, as usual, often plays a big part in workplace politics. I have said it before and I will repeat myself even if it
means being considered senile. Do not participate in politics except as required for survival. Backstabbing, under-
mining, and sidestepping projects all take a great deal of
energy and produces nothing but resentment and a desire to
retaliate. I have listened to and participated in all the little
games used to get ahead in corporate America, only to realize that honesty and hard work pay much larger dividends.

A senior engineer at a plant with severe ergonomic problems told me to my face that if it weren't for OSHA, I would
have to get a real job. It seems his ego was telling him I
shouldn't be on his turf. Interestingly, his ego did not point
out the fact that, due to his ineffectiveness as an engineering
manager, so many people had been injured throughout the
corporation that OSHA finally noticed and red-flagged several dangerous operations. I later learned that he got his ego-
tistical fix by playing compliance word games with the EPA

and OSHA. He felt better "outwitting" government agencies at their own game. It never once occurred to him that many employees were paying exorbitant life-and-limb admission fees to his non-compliance chess games.

Do not confuse being tactful, polite, and understanding with playing politics. I have known several crude and abrasive individuals who justify their behavior by claiming that they're not playing politics. Their success rate at implementing anything useful was just about nonexistent. The key is successful project implementation. Is it happening or not?

Is it happening or not? Interesting question, as it is very difficult for most engineers and managers to answer honestly. If you subscribe to the Panic Management Theory of project management and frequently run screaming from inferred and imagined problems, you cannot afford to look too closely at the results. Dealing in facts can be very punishing. However, it softens the blows to your ego and career if you can learn to look sideways at reality.

Exercising selective memory loss when embarrassing aspects of projects come up in conversation is another common practice. I know of one up-and-coming automotive division that takes this attitude: "Keep it quiet and no one will know just how bad we are at building cars." It is a standard industry joke to relate the television commercials big-name ad agencies develop to the reality only known to insiders in the plant. Only an idiot would have bought these cars if they had known the quality problems that continued to plague the division well past a normal start-up period. I don't even want to know just how widespread this attitude is in industry today.

The problem of the "Plaque on the Wall Syndrome" is almost as bad and maybe worse, in its own way, than any trap or pitfall I have already mentioned. Generally, I have only encountered this phenomenon in its most blatant form in privately held companies. Their owners want recognition in the media as innovators and followers of the latest guru-driven alphabet soup. Ego stroking notwithstanding, I see no use in pseudo-implementing (I'd love to write "half-ass implementing," but I'm not sure how well that would go over) a technique just be regarded as a forward-thinking, paradigm-shifting industry leader. The first hint of hard, sweaty work generally turns the stomachs of these big guys in the front office. Employees who are told repeatedly that deep pockets do not exist at XYZ company (and the myth of deep pockets is a very serious problem) become disillusioned as they gaze upon the brass and walnut plaques. Why? They realize how much money was spent for each plaque and are painfully aware of the problems that are never addressed on the manufacturing floor or in the front office. Most have learned to laugh and shrug it off as they go about their usual business. However, when they are asked their honest opinions, it is readily apparent that bitterness is lurking just below the surface.

If an owner wants a new technique to work, the managers must buy into it and work hard at making it successful. Unfortunately the winds can change without warning and suddenly the chase for new plaque is on. Meanwhile, the old plaque will start gathering dust. Managers not quick enough to jump bandwagons can find themselves unemployed. It follows, quite reasonably, that managers refuse to commit to new techniques, and so they fail. No problem— a new technique is on its way. The emphasis now is surviving in a given position, not successfully implementing a plan and making things better.

An upper-level manager of a very large, privately owned brewery was giving an impassioned presentation on Total Quality Management (TQM) to a group of engineering students. I raised the question about manager commitment and how it was handled. His snappy response was that the offending managers are taken out and shot. After the laughter died down, I asked the question again. The sobering answer? Upper management treated the projects the same way as their lower level counterparts and for the same reasons. No real solutions had ever been found. Effectiveness was minimal, but better than not attempting any change. If you think the line workers in that corporation didn't have an opinion on that philosophy, think again.

I now have a couple of classic traps and pitfalls to discuss. Always address the problem and not the symptom. Always follow up on a project until it is successfully completed or left for dead.

A friend of mine was asked to troubleshoot his grandmother's electric furnace. It kept blowing a fuse in the middle of the night (when the heat was needed most), and the monthly electric bills were tremendously high. Rather than calling an HVAC contractor to look at the furnace, he decided that repairing the fireplace and installing an insert would heat the house more effectively. He believed the house being cold was the problem that needed to be addressed. Several hundreds of dollars and a half a cord of wood later, the house was still cold. The next idea was obtaining a special permit for installing a propane tank next to the house and getting a brand new heat pump. It was also suggested that a branch line from the main city gas line be run and a new gas furnace be installed.

At this point, I was asked my opinion. I suggested calling a repairman and then considering the cost payback of an expensive heat pump or gas furnace after his report. The repairman installed a properly sized fuse, tightened a loose connection on the electric heater, and the house became warm. I believed the problem was that the furnace was not working properly and that the cold house was a symptom of that problem. A lot of time, money, and cold extremities could have been saved if the problem, and not the symptoms, had been addressed from the outset.

My last trap and pitfall follows from the above example. Although the wrong characteristic was being addressed, the problem was not dropped or abandoned. The follow-through on all aspects of the project ultimately corrected the problem and an intolerable circumstance was made livable again. I would suggest that you treat every project you undertake as if proper problem identification and follow-through are the only things that will keep your grandmother warm on a freezing winter's night.

# CHAPTER 8

# Selected Alphabet Soups

Corporations, like our mothers before them,
are buying into and ladling out
industrial versions of alphabet soup
—well-packaged cans that promise
corporate health and well being.

SOUPS ARE THOSE SIMMERING CONCOCTIONS that contain all the basic ingredients we know and love. A little broth, vegetables, seasonings, a thickening agent, and meat—if you are so inclined. (Technically, meat as a component constitutes a stew, but my brother is the chef, not me.) *Joy of Cooking,* my guide to gastronomic delights, lists over 160 types of soups and soup components. (Yes, I know, stews are a completely different section numbering about 30 items, but I've already addressed that issue.)

Amazingly, among the minestrone, bouillabaisse, chowder, gumbo, and miso soups, there is no "alphabet" soup. I don't mean to desecrate an American institution, but perhaps all those red-and-white cans of alphabet soup we inhaled as kids weren't all they're cracked up to be. Perhaps those wonderful little letters were merely a marketing ploy to convince us that the can's contents weren't ordinary, everyday soup stock items. Instead, those tiny letters somehow transformed carrots and broth into otherworldly items that could lay waste to any diseases our mothers suspected were lying in wait to pounce on our young and vulnerable developing bodies. We fully believed we weren't eating vegetables. We were eating alphabet soup, the cure-all, end-all of foods. Why, it was good for everything and anything, including sating hunger. Now it seems that corporations, like our mothers before them, are buying into and ladling out industrial versions of alphabet soup—well-packaged cans that promise corporate health and well being.

Take, for instance, my favorite industrial alphabet soups labeled TQM (Total Quality Management), ISO 9000 (International Quality Standards), and BPR (Business Process Re-engineering). Who hasn't heard of these? They and a hundred other acronyms claim to spell out the future of manufacturing and business. It is so bad that companies are starting to create their own internal custom soups. QFD and STEM are legitimate examples. Of course, there are the companies that are adopting JIC (Just-In-Case) manufacturing as a popular spin-off of JIT. Give me a break, or give me a clue! The packages are pretty. The descriptions rate inclusion in a reference dictionary and the gurus all speak with thunder from the mountain tops. So what. Look inside the package and past the marketing. You'll find those tiny little letters trying to convince us once again that the package is mystical. Ignore the carrots behind the curtain, if you please.

TQM. Total Quality Management. I admit it. It was two or three years after I first heard about it before I discovered that it has little or nothing to do with quality. TQM is simply a management technique that used to be called Quality Circles. Quality Circles, by the way, used to be called "talking to your employees." If you know how to manage projects, you understand how to delegate, and you believe in your employees and support their ideas, you are doing TQM. For all those squealing gurus that are about to shout that I obviously don't understand the finer points of TQM, save your breath. You obviously don't understand the finer attributes of the basic building blocks hundreds, if not thousands, of engineers and managers employ to make business and industry work every day without the benefit of those tiny white letters. Besides, TQM is, as of this writing, out of vogue and TQP (Total Quality Process) has taken its place. It must be better; it is new. (Big smile, star-twinkle glinting off the diamond embedded front tooth of yet another slick consultant who wants your money and to waste your time.)

TQM is simply a management technique that used to be called Quality Circles. Quality Circles, by the way, used to be called "talking to your employees." If you know how to manage projects, you understand how to delegate, and you believe in your employees and support their ideas, you are doing TQM.

ISO 9000. Wow. The international standards. Please, let me drop to my knees in wonder at this latest and greatest wisdom that will save manufacturing as we know it. Talk about a really cool marketing ploy that throws money to gurus who are laughing all the way to the bank. Think of all the opportunities for heaving thunderbolts at unsuspecting and innocent manufacturers until they bow their heads in supplication.

Since I'm a consultant (it really isn't a four letter word, although I have heard it used as such), I felt it would be a great idea to be a certified ISO 9000 auditor. That way I'd be in a better position to help my clients implement ISO 9000 and gain registration. What a shock. I couldn't find any information on what ISO 9000 was or what it took to become certified. Instead, I learned that I needed to spend time on an approved team to gain auditing experience. After an apprentice period, I would be given the mantle of guru auditor, assuming I was good and met the predetermined criteria. Later, I could work my way up to Team Leader and be able to throw thunderbolts. Way cool. After carefully perusing the material, however, I still had no idea what being ISO 9000 certified meant. Interestingly, most of my clients and friends investigating certification for their facilities couldn't find out either.

ISO 9000 is simply a checklist of items that must be addressed and documented before one's company is knighted and allowed to play on the ISO 9000 playground.

Our collective mistake was once again caused by those little white letters floating around in front of our eyes. Years later, we have all learned that ISO 9000 is simply a checklist of items that must be addressed and documented before one's company is knighted and allowed to play on the ISO 9000 playground. The items on the checklist are the standard carrots and broth: quality programs, employee training, etc. There is nothing new or improved on the list, just the same old items that college management and engineering texts teach you. If you don't have a college degree, 10 or 20 years of on-the-job experience will teach you the same things—if you take your job seriously. The big catch is that special documentation must be implemented and maintained to prove to the specially trained and approved auditor that every item on the list is in compliance.

Did I fail to mention that the ISO 9000 standard does not spell out or specifically detail how the items on the checklist are to be accomplished? That's right. The auditor comes in and determines if you are in compliance with the clearly delineated documentation procedures and then verifies that the items specified are in place. If the auditor doesn't like what he or she sees, your hand is slapped and your certification denied. After a suitable waiting period you are again graced by the auditor's presence, and the process starts over again. The audit is not cheap. But, here's the good news: if you are clever and spend the bankful of money needed to hire an army of documentation personnel to complete the paperwork correctly, you can become certified. The only catch is that the audit doesn't differentiate between programs that successfully address your facility's requirements and those that merely fill in the boxes on the ISO 9000 forms. It also has to be repeated about every six months for just a few thousand more dollars.

The number of books written on how to "interpret" the standards indicates just how nebulous most of the concepts are when it comes to implementing them in the real world. Honestly, any standards that are so generic they apply to all industries across the board can't possibly address specifics. And if you are not addressing specifics, you are not accomplishing anything useful.

An engineer who works for one of my clients as their internal ISO standards interpreter and implementor has a fairly low overall opinion of the standards. In a moment of honesty and clarity, he summed it up quite succinctly: "ISO 9000 promotes bad management." Over lunch the other day he elaborated and tried to put a good face on ISO. "ISO is an excellent documentation checklist." Engineers and managers I know repeat this as the sole positive item floating in a sea of frustration and anger caused by trying to grasp the golden ring of ISO certification for their companies. Unfortunately, the good face he was painting on ISO quickly turned into a grimace of distress. He began relating his experiences and those of other companies pursuing ISO trophies: incredible costs, mountains of paperwork, lots of generic theory, and absolutely no practical or applied information.

I have a copy of the standards and thus will share a tiny slice of ISO wisdom. "Section 4.15.2 Handling. The supplier shall provide methods of handling product that prevent damage or deterioration." Hel-ll-oo-oo? I think I will manufacture a product and then intentionally damage it in handling, while pretending it is not a problem. I won't be ISO-certified, but that's okay, because my clients all buy damaged product now. Even if I assume that I experienced an epiphany at this section, how do I go from the standard's generic statement to an applied manifestation on my shop floor? If I don't recognize damaged or deteriorated product now, how will I determine that handling methods need to be modified to prevent damage? If I recognize the damage because of the ISO standard (okay, so it's a stretch, bear with me) how do I know what to do to modify the handling methods? Perhaps the answer lies in my education, experience, or personality. You got it: ISO may point out the problems, but OFE&M will provide the solutions.

In theory, it is assumed that by filling in those boxes, a company will be worthy of certification as a robust, forward-thinking, and successful example of the best in their particular industry. Acceptance into the ISO 9000 clan is no less than a sanctification. Unfortunately, the reality is that many certified companies are merely very good at completing forms and keeping records. Here's a classic real world example: a big, important client told a manufacturer of electronic solenoids that they must be certified or the client would stop purchasing from them. After a lengthy review of the costs and effort, the solenoid manufacturer decided to pass. Did they lose the business? Not hardly. As the world's leading manufacturer of quality solenoids, they realized something that many companies are now learning. If you produce a quality product, certification is not necessary because the product will stand on its own. The client, who was certified, purchased the solenoids through a third-party warehouse. Their certification allowed a different set of standards when purchasing items from a distributor than from a manufacturer. Of course, they had to pay an additional markup and lead times were extended.

ISO has a place in manufacturing. It is a basic documentation tool—not an end in itself. As such, use the checklist for those areas needing improvement and ignore the certification process and its associated costs. Several of my clients are doing just that. So much for the miracle of ISO.

BPR. Business process reengineering. Oh, boy. Those three little letters will ruin a good soup faster than a heaping handful of salt. Actually BPR does make a good stew. General BPR practice involves raping a company to the bone without discriminating between muscle or fat. Throwing away the vital meat of a company creates a stew out of that seemingly innocent alphabet soup. (You knew I referenced soup vs. stew for a reason, didn't you?)

BPR is a quick fix for the symptoms that plague a company. Almost everyone I have talked to about their BPR experiences relates the same horror stories of trying desperately to address the symptoms while ignoring the basic problems. Cutting to the bone and restructuring from the ground up buys a company some time, but the real problems will still catch up eventually if they are not addressed. How can a company or process that has been successful in the past suddenly turn unprofitable? Market shift, technological or regulatory changes, and worldwide competition are possible reasons, but rarely have I seen justification for the wholesale slaughter and sacrifice of employees to restructuring. As a manager, you must be aware of the effect watching one of these reengineering projects in action will have on your remaining employees' morale. BPR reminds me of the cartoon of the little boy who would be king. He is sitting on the throne pouting as his royal advisor explains, "You can't say 'on with his head' after you say 'off with his head.'" Many times it is too late to recover the vital skills and knowledge lost after BPR cuts its swath through a company.

The latest twist to this brand of alphabet soup is really sad: reapplication. This is allowing those workers targeted by the reengineering process the opportunity to reapply for their old positions at a lower salary with loss of benefits. It is a guaranteed morale buster.

I'm afraid the alphabet soup we ate as kids severely damaged that part of our brains that allows us to differentiate between basic truths and really neat, totally believable marketing schemes. These are schemes designed to make us believe we are getting something more—something special, when actually we are getting basic components packaged in a warm, fuzzy wrapper.

A potential client requested my attendance at three separate fact-gathering team meetings to determine if I had specific knowledge that addressed their quality concerns. I used OFE&M as a developing philosophy long before I realized that I needed to give it a fancy title. Therefore, I fell woefully short of their expectations as a consultant. I didn't know, let alone use, the most current catch phrases. Unfortunately, the team leader was incredibly impressed with the other interviewee's knowledge of current marketing jargon.

I blew it big time. My personality just is not suited to playing political games. I bluntly refused to buy into their chase for a really "neat-sounding" technique that would address their quality concerns. I disappointed the team leader so much that I couldn't even get any answers about my bid from anyone. Finally, the secretary informed me that she was relatively certain I hadn't gotten the contract, but she wasn't sure. I swallowed my disappointment and went in search of greener pastures.

The company hired a husband and wife team formerly employed by a large telephone company. They had done lots of internal training in really neat sounding techniques. My thoughts, mirrored by the team members that wanted to hire me, were, "If they are so good, why are they no longer employed by that company?" This naturally progressed to, "How are they going to address any serious problems with these nifty phrases?"

Two years later, I got a phone call during a relaxed summer evening at home. It was the president of the company that had rejected me. It seems he wanted to know if I was still in business and if I would be interested in trying to help them solve their quality problems.

I told him I had the classic case of good news and bad news. The good news was that I was still in business. The bad news was that my rates had doubled since we last talked. He was very congenial and offered to set up an appointment at my convenience. I was suspicious because no other president of any size business has ever been that polite and straightforward with me. Our meeting turned out to be enlightening, and I am proud to say I learned a lot from it.

It seems the "catch phrase team" had no clue about how to turn a manufacturing plant's quality program around. Everyone across the board felt really "good" about quality and believed that it was an important component of their daily tasks. However, no one really understood how to actually do anything about the quality problems the plant was still experiencing. An empty warm fuzzy does not a good part make.

My approach was very basic and hands-on (as it always is).

After two weeks of analysis and one four-hour training session for key people, the quality program had a solid framework. I designed it so that internal personnel could flesh it out and make it work. The new project manager remarked that I should write a book on "nuts and bolts quality programs." The half-dozen high-dollar seminars he had attended had not yielded as much applicable information as my analysis and training. I want to thank him now for his comments.

OFE&M was born then and there, although I didn't know it at the time. I realized that my approach to quality programs was the same as my approach to all engineering and management problems. I have always taken what I know and applied it with as little fear and as much confidence as I could muster. "Nuts and Bolts Engineering and Management" didn't seem as catchy as "Old-Fashioned Engineering and Management," so although it was at his suggestion this book came into being, I felt I should take some creative license with the title.

I realized that a book on engineering and management basics needed to have a hook if it was to sell a million copies, allowing me to retire at a ridiculously young age. Hence, OFE&M. It's still the same philosophy, but admit it, the alphabet soup title originally attracted you and caught your attention. In fact, I hope you have bought this book before reaching this section and realizing that it is really about the nuts and bolts of engineering and management. (Actually, since I admitted as much in the preface, I do not harbor any feelings of guilt.)

# CHAPTER 9

# The Bottom Line

Creative title notwithstanding,
OFE&M is simply a reminder to use what you
know. Think of it as a "Post-It Note"
giving you permission to apply your skills
and knowledge creatively and innovatively
without fear of being labeled "uninformed."
After all, you are practicing OFE&M.

OLD-FASHIONED, IN THIS TECHNOLOGICAL AGE, HAS GAINED a negative connotation as an adjective. It seems that, if one already has knowledge, it is considered outdated and obsolete. In some cases this is true, but old-fashioned used as a philosophy will never become useless. Imagination and a willingness to embrace new information that has true value will always be in vogue. The hardest part of OFE&M is taking responsibility for one's actions and deciding which path is worth pursuing.

Sifting through the chaff of "new" techniques and methodologies constantly bombarding us can be rather daunting. Seeking to improve areas that have been identified as needing more polish is not as easy as signing up for every seminar that professes to cure what ails you or your company. Taking charge of your career and developing awareness of what is happening around you so that you can take positive action is difficult. It is much easier to daydream through a lecture and shelve the notebook. Making responsible decisions is hard work, as is acting on them, but it has its rewards.

I would hope that as readers of this book, you come away with an attitude similar to that of those who attend my lectures on OFE&M. The most popular comment is "I didn't really learn anything new today, but I see now where I can successfully apply what I know in many more situations than I previously thought."

OFE&M isn't new, or improved, but it is a time-tested philosophy that is effective.

# About the Author

MARK A. OUSNAMER, M.S., P.E., is the owner of Industrial Engineering Services, located in Kansas City, Missouri. The firm offers a full line of manufacturing- and service-related engineering services to government, municipal, industrial, commercial, medical, and institutional clients.

Mark has more than thirteen years of hands-on experience balanced with teaching and lab research in the following areas: industrial engineering, ergonomics, environmental engineering, fluid dynamics, structural engineering, concurrent engineering, information systems design, manufacturing engineering, and forensic engineering. A senior member of IIE and the Society of Manufacturing Engineers, he received a Masters of Science degree in industrial engineering from the University of Missouri, and a Bachelor of Science degree in industrial engineering from the University of Arkansas.

# About EMP

ENGINEERING & MANAGEMENT PRESS (EMP) is the award-winning book publishing division of the Institute of Industrial Engineers (IIE). EMP was awarded the 1996 *Association Trends* Publishing Award for best book/manual in the soft cover category for *Manufacturing and the Internet* by Richard Mathieu. EMP was also one of four finalists for the 1996 Literary Marketplace *Corporate Achievement Award* in the Professional Category.

EMP was founded in 1981 as Industrial Engineering & Management Press (IE&MP). In 1995, IE&MP was reengineered as Engineering & Management Press. As both IE&MP and EMP, the press has a history of publishing successful titles, such as *Toyota Production Systems, Winning Manufacturing, Managing Quality in America's Most Admired Companies,* and *Beyond the Basics of Reengineering.*

Persons interested in submitting manuscripts to the press should contact Forsyth Alexander, Book Editor, EMP, 25 Technology Park, Norcross, GA 30092.

# About IIE

FOUNDED IN 1948, THE INSTITUTE of Industrial Engineers (IIE) is comprised of more than 25,000 members throughout the United States and 89 other countries. IIE is the only international, nonprofit professional society dedicated to advancing the technical and managerial excellence of industrial engineers and all individuals involved in improving overall quality and productivity. IIE is committed to providing timely information about the profession to its membership, to professionals who practice industrial engineering skills, and to the general public.

IIE provides continuing education opportunities to members to keep them current on the latest technologies and systems that contribute to career advancement. The Institute provides products and services to aid in this endeavor, including professional magazines, journals, books, conferences, and seminars. IIE is constantly working to be the best available resource for information about the industrial engineering profession.

For more information about membership in IIE, please contact IIE Member and Customer Service at 800-494-0460 or 770-449-0460 or cs.@www.iienet.org.

# Industrial Management

FOR NEARLY 40 YEARS, *Industrial Management* magazine has been serving the needs of business managers concerned with improving processes, productivity, and quality. In every industry—from manufacturing to service to government, *Industrial Management* provides in-depth and insightful coverage of topics like these:

> *business process reengineering, collective bargaining, concurrent engineering, cycle time reduction, forecasting, human resources strategies, industrial organization, Kaizen, Kanban material acquisition, labor relations, maintenance, management of technology, manufacturing management, operations improvement, organizational behavior, performance measurement, planning, production, productivity, quality, rapid prototyping, scheduling, service, strategic planning, theory of constraints, total quality management, union relations, and work teams.*

Every other month, *Industrial Management* provides case studies, practical advice, and hands-on techniques. Subscribers learn from the experiences of others how to avoid unnecessary and potentially costly mistakes—and keep their careers in management on the right track. Each issue contains articles that present to readers both the how and the why of management techniques before they try them out in the workplace.

*Industrial Management*, a bimonthly magazine, is the official publication of the Society for Engineering and Management Systems (SEMS) of the Institute of Industrial Engineers (IIE). Current subscription rates: $39.00 per year (within the United States); $50.00 per year (outside the U.S.). Prices are subject to change. Discounts are available to members of IIE. There is an additional charge for airmail delivery. To subscribe or to receive more information, contact IIE Member and Customer Service at 800-494-0460 or 770-449-0460 or visit our web site at www.iienet.org.

# Other Books from EMP

TEAMBUILDING AND
TOTAL QUALITY
*A Guidebook To TQM Success*
    by GENE MILAS
    hardcover, 1997
    ISBN 0-89806-173-3
    order code: MILTQM
    list price $29.95

LESSONS TO BE LEARNED
JUST IN TIME
    by JAMES J. CAMMARANO
    hardcover, 1997
    ISBN 0-89806-162-7
    order code: LESSON
    list price $34.95

SIMULATION MADE EASY
*A Manager's Guide*
    by CHARLES HARRELL, PH.D.,
    and KERIM TUMAY
    311 pages, hardcover, 1995
    ISBN 0-89806-136-9
    order code: SIMSFY
    list price $50.00

FACILITIES AND
WORKPLACE DESIGN
*An Illustrated Guide*
    by QUARTERMAN LEE with ARILD ENG
    AMUNDSEN, WILLIAM NELSON, and
    HERBERT TUTTLE
    232 pages, softcover, 1997
    ISBN 0-89806-166-0
    order code: FACDGN
    list price $25.00

DESIGN OF EXPERIMENTS FOR
PROCESS IMPROVEMENT
AND QUALITY ASSURANCE
    by ROBERT F. BREWER, P.E.
    280 pages, softcover, 1996
    ISBN 0-89806-165-2
    order code: BREWER   list price $25.00

ESSENTIAL CAREER SKILLS
FOR ENGINEERS
    by SHAHAB SAEED, P.E.,
    and KEITH JOHNSON, P.E.
    112 pages, softcover, 1995
    ISBN 0-89806-142-3
    order code: BUSSKI  list price $25.00

WORK SIMPLIFICATION
*An Analyst's Handbook*
    by PIERRE THÉRIAULT
    200 pages, hardcover, 1996
    ISBN 0-89806-163-6
    order code: THERIA  list price $25.00

BY WHAT METHOD?
    by D. SCOTT SINK, PH.D.,
    and WILLIAM T. MORRIS, P.E.
    364 pages, softcover, 1995
    ISBN 0-89806-141-5
    order code: BWM752  list price $30.00

TOYOTA PRODUCTION SYSTEM
*An Integrated Approach to Just-In-Time*
    by YASUHIRO MONDEN, PH.D.
    425 pages, hardcover, 1993
    ISBN 0-89806-129-6
    order code: NWTYPS  list price $53.95

## To order books from EMP

or to request a free catalog of EMP's titles, please call IIE Member & Customer Service
at 800-494-0460 or 770-449-0460. We also invite you to visit us
on IIE's web site at http://www.iienet.org.